PHYSICS RESEARCH AND TECHNOLOGY

GADOLINIUM FOILS AS CONVERTERS OF THERMAL NEUTRONS IN DETECTORS OF NUCLEAR RADIATION

PHYSICS RESEARCH AND TECHNOLOGY

Additional books in this series can be found on Nova's website under the Series tab.

Additional E-books in this series can be found on Nova's website under the E-books tab.

PHYSICS RESEARCH AND TECHNOLOGY

GADOLINIUM FOILS AS CONVERTERS OF THERMAL NEUTRONS IN DETECTORS OF NUCLEAR RADIATION

D. A. ABDUSHUKUROV

Nova Science Publishers, Inc.
New York

Copyright © 2010 by Nova Science Publishers, Inc.

All rights reserved. No part of this book may be reproduced, stored in a retrieval system or transmitted in any form or by any means: electronic, electrostatic, magnetic, tape, mechanical photocopying, recording or otherwise without the written permission of the Publisher.

For permission to use material from this book please contact us:
Telephone 631-231-7269; Fax 631-231-8175
Web Site: http://www.novapublishers.com

NOTICE TO THE READER

The Publisher has taken reasonable care in the preparation of this book, but makes no expressed or implied warranty of any kind and assumes no responsibility for any errors or omissions. No liability is assumed for incidental or consequential damages in connection with or arising out of information contained in this book. The Publisher shall not be liable for any special, consequential, or exemplary damages resulting, in whole or in part, from the readers' use of, or reliance upon, this material.

Independent verification should be sought for any data, advice or recommendations contained in this book. In addition, no responsibility is assumed by the publisher for any injury and/or damage to persons or property arising from any methods, products, instructions, ideas or otherwise contained in this publication.

This publication is designed to provide accurate and authoritative information with regard to the subject matter covered herein. It is sold with the clear understanding that the Publisher is not engaged in rendering legal or any other professional services. If legal or any other expert assistance is required, the services of a competent person should be sought. FROM A DECLARATION OF PARTICIPANTS JOINTLY ADOPTED BY A COMMITTEE OF THE AMERICAN BAR ASSOCIATION AND A COMMITTEE OF PUBLISHERS.

LIBRARY OF CONGRESS CATALOGING-IN-PUBLICATION DATA
Abdushukurov, D. A.
Gadolinium foils as converters of thermal neutrons in detectors of nuclear
radiation / author, D.A. Abdushukurov.
p. cm.
Includes index.
 ISBN 978-1-61728-676-6 (softcover)
1. Neutron counters--Materials. 2. Gadolinium. 3. Metal foils. I.
Title.
QC787.C6A155 2010
539.7'7--dc22-2010025428

Published by Nova Science Publishers, Inc. New York

CONTENTS

Preface		vii
Chapter 1	Introduction	1
Chapter 2	Mathematical Modeling of Converter Perfomances	5
Chapter 3	Position Sensitive Detectors of Thermal Neutrons with Gadolinum Converters	71
Conclusions		111
References		115
Index		121

PREFACE

Converters of neutron radiation play a determining role in the development of detectors of these radiations. They determine the basic characteristics of detectors: the efficiency of registration, energy, time and spatial resolution.

Among solid-state converters on the basis of gadolinium and its 157 isotopes are especially allocated, possessing an abnormal high cross-section of interaction with thermal neutrons.

In the article, theoretical bases of registration of neutron radiation by converters from gadolinium are considered. The efficiency of converters is the product of three variables. These are the following: Probabilities of capture of thermal neutrons by nucleus; Probabilities of creation of the secondary charged particles, in our case of internal conversion and Auger electrons; Probabilities of escape created electrons from the material of the converter.

Model calculations of registration efficiency of thermal neutrons by the foil converters made from natural gadolinium and its 157 isotope described. Processes of neutron absorption in the material of a converter and the probability of secondary electron escapes are examined. Calculation is made for converters with the various thicknesses, and other parameters of converters. It was chosen the most optimal converter thicknesses.

The contribution low-energy Auger electrons radiated from L- subshell with the energy 4.84 keV and M-subshell with the energy 0.97 keV on efficiency of converters are lead. These electrons have rather small free path length in gadolinium; these are 0.3 microns (4.84 keV) and 0.04 microns (0.97 keV). But their contribution to become essential at use of converters from 157 gadolinium isotopes as the length of free path of neutrons in them does not exceed 2-3 microns, and this length is to become comparable with length of path electrons.

The estimation of contribution of X-ray and low-energy gamma-quanta is absorbed directly in the converter and resulting in occurrence of secondary electrons. In case of the account of the contribution of electrons formed by X-ray quanta, the efficiency is increased a little, but their contribution is no more than by 1%.

Calculations of complex converters representing a set thin gadolinium foil located on both sides of supporting kapton foils and calculations of complex converters representing a set thin drilled with the fine step foils located one over other in a gas volume are lead.

Examples of development of detectors of neutrons based on gadolinium converters described.

Chapter 1

INTRODUCTION

Mechanisms for detecting neutrons in matter are based on indirect methods. Neutrons are neutral; they do not interact directly with the electrons in matter. The neutrons can cause a nuclear reaction. The process of neutron detection begins when neutrons, interacting with various nuclei, initiate the release of one or more secondary particles. The products from these reactions, such as protons, alpha particles, gamma rays, electrons and fission fragments, can initiate the detection process. For detection of neutrally charged neutrons, it is necessary to use converters of neutron radiation. These converters will converse radiations of neutrons to the charged radiation, which can be further detected with manifold detectors.

For an optimal choice of the converter foil material, the following properties should be considered:

- A large neutron-capture cross-section needed in order to achieve high detection efficiency with minimum foil thickness.
- The range of the neutron-induced charged primary particles should be large compared with the convertor foil thickness.
- The converter material should possess small cross-section of activation. Daughter isotopes occurring as result of activation should not create the significant background.
- The γ-ray sensitivity of the material should be low.

At the present time, gaseous and solid-state converters are used. Thus, several kinds of nuclear reactions are used in these converters. The most widely known are ^{3}He and ^{10}BF$_3$ based converters [1]. Gaseous converters are used in the wide class of gas detectors. Solid-state converters are applied in gaseous, semiconductor and in scintillation detectors.

In the development of gaseous detectors of thermal neutrons, the highest efficiency of registration up to 80% has been reached to receive with the use of gas mixes on the basis of Helium - 3 (^{3}He) at pressure up to 15 atm. The use of the high pressure is necessary in order to increase the efficiency of registration and improvement of the spatial resolution. The increase of pressure dictates the additional requirements to the mechanical designs and first of all to the thickness of the entrance window. At the same time, the high price of the gas and fluidity of ^{3}He results in sharp rise in price of detectors and their operation.

Converters of neutron radiation play a determining role at designing detectors of neutron radiation. They define the key parameters of detectors, such as the efficiency, the spatial resolution and so on. Among the solid-state converters of thermal neutrons, the highest efficiency of registration has been received using the gadolinium-based converters, and especially, its 157 isotope. As a result of radiating capture of thermal neutrons by nucleus of gadolinium, electrons of internal conversion and Auger electrons are radiated. Another solid-state converter is rarely used in detectors of neutrons, which is first of all connected with the low efficiency of registration. Therefore, the efficiency of registration of detectors based on ^{10}B and ^{6}Li usually do not exceed 3-4% and 1%, respectively [2,3].

The idea of using thin foils to convert neutrons radiation into charged particles and count the conversion products in a solid-state detector was originally proposed by Feigl and Rauch [4,5]. They employed natural Gd and pure ^{157}Gd convertor foils and measured the escape probabilities and the energy distributions of the escaping electrons with Si surface barrier detectors.

In the paper by A.P.Jeavons et al. [6], the efficiency of registration of thermal neutrons by the converters constructed based on gadolinium foils has been simulated. The following conditions were considered at the modeling: the neutron beam (two fixed energies) perpendicularly fell onto the converting foils of various thicknesses; the output of the secondary electrons has been calculated both in lobby and back hemispheres. As the result, the detector on the basis of the multiwire proportional chamber with the gadolinium converter has been developed. Nevertheless, such a detector did not have enough good spatial resolution, about 10 mm. In order to improve the spatial resolution, the same authors have offered to use the converter - collimator, which was the sandwich created from lead foils with the fiberglass layers and drilled with the fine step (Jeavons converter). This converter was nestled to the gadolinium foil and thus limited electron runs in the gas, simultaneously serving as the emitter of the secondary electrons. It has helped to improve the spatial resolution up to 2 mm, but the spatial resolution has appeared modulated with step of apertures, besides the efficiency of registration has sharply fallen.

Further, in G. Charpak group [7], by the development of multistep avalanche chambers with gadolinium converters, it was managed to improve sharply the spatial resolution of detectors (up to 1 mm) without use of Jeavons converter [8].

In the paper by D.A.Abdushukurov et.al. [9], modeling of the efficiency of registration of thermal neutrons by the gadolinium foils has been conducted. In contrast to earlier works, the incidence angle of neutron beam was not a constant value (90 degrees) and varied from 1 up to 90 degrees. As a result, the conclusion about the increase of efficiency of registration of neutrons at the small angles (up to 10 degrees) between the converter and beam of incidence neutrons has been done. It is connected with the increase of neutron path length in the body of the converter at conserving of effective thickness of material for the output of secondary electrons. At the disposition of two detectors on the different sides of the converter, the efficiency of registration of thermal neutrons can be increased up to 60%. In the carried out calculations, we took into account only electrons with the energies higher than 29 keV. Thus, only electrons of internal conversion and Auger electrons, radiated from K-shell, with the energy 34.9 keV were taken into account [10]. The minimal free path length of electrons with energy 29 keV in gadolinium makes 4.7 microns. The similar choice of energies of electrons was made not only by us, but also by other researchers.

Recently, calculations of influence of low-energy secondary electrons radiated from gadolinium foils during radiating capture of thermal neutrons were carried out. These are Auger electrons radiated from a L-subshell with the energy 4.84 keV and electrons from M-subshell with energy 0.97 keV. These electrons have rather small free path length in gadolinium; these are 0.3 microns (4.84 keV) and 0.04 microns (0.97 keV), thus accordingly make average free path length of these electrons 0,1 and 0,015 microns. During calculations, it was found out that their contribution becomes essential at use of converters made from 157 isotope of gadolinium as the length of free path of neutrons in them does not exceed 2-3 microns, and this length becomes comparable with the free path length of electrons. Results of modeling calculations within the limits of errors have coincided with the experimental data and published in [11]. The good consent of the calculated data with the experimental ones testifies to correctness of the chosen models and theoretical preconditions.

In the literature and the tabulated data, there are no data for the free path length of electrons with the energy less than 10 keV. All available data for electrons in various materials were received from the Bethe-Bloch theory. The lower border of applicability of this theory is the energy of 10 keV. The theory is constructed on the assumption that the charged particles have continuous losses of energy at their motion in materials by two mechanisms. These are losses of energy on irradiations and direct collisions. For electron energies less than 10 keV the new mechanism added, it is the probability of capture of free electrons by the atom subshells having vacancies. Also practically, there are no data on the quantity of low-energy Auger electrons. Available data differs with an error more than 100%. During realization of calculations, it is necessary to compare received results with the experimental data. It will allow estimating the correctness of theoretical preconditions and modeling representations.

The thin layer converters will allow creating new types of detectors of thermal neutrons having the improved characteristics. Therefore, it will be possible to create detectors with pico-second time resolution for time-resolved experiments; also an improvement of the spatial resolution can expected due to the reduction of thickness and parallax of detectors. It is possible to specify separately an opportunity of miniaturization of detectors that claimed for many applications.

Computer modeling allows to prospect, without special material inputs of the most suitable configuration of converters, to carry out search of optimum geometrical ratio. It in turn allows refusing realization of superfluous development and researches.

In present time to position-sensitive detection of thermal neutron radiation widely applied various detectors with solid-state converters such as normal-pressure multistep avalanche chambers (MSAC), low-pressure multistep avalanche chambers (LMSAC), microchannel plates, thin layer scintillators. New types of detectors are offered and developed, such as hybrid low-pressure micro-strip gas chamber (MSGC), position-sensitive silicon detectors (Gd Si PD), resistive plate chambers (RPCs), imaging plate neutron detectors (IP-NDs), liquid scintillator into a capillary plate and so on.

Chapter 2

MATHEMATICAL MODELING OF CONVERTER PERFORMANCES

2.1. THEORETICAL BASES

In the current chapter, influence of various parameters of converters on their efficiency are viewed. Efficiency in our understanding is the ration of the electrons that have departed from converters, to total number of falling neutrons on the converter. The efficiency of converters is the product of three variables. These are the following:

- Probabilities of thermal neutrons capture by nucleus of the converter, which depends on its thickness, and the cross-section of interaction. The cross-section of interaction, in turn, depends on isotope composition of the converter and energy or wavelength of neutron.
- Probabilities of creation of the secondary charged particles, in our case of internal conversion and Auger electrons.
- Probabilities of escape of the created electrons from the material of the converter, which depends on free path length of electrons in a converter material and geometry of emission.

Energies of the secondary electrons are discrete values and are formed with the certain probability. At calculation of probability of escape, it is necessary to take into account emission of each electron with its characteristic energy and its weight factor (probability of formation). In that specific case, efficiency can be described by the following expression

$$E_i = P_n(\sigma L_j) * D_e (E_i N_i) * H_e(Re_i\Theta_i) \qquad (1)$$

where E_i is the efficiency of registration of the absorbed with the probability $P_n(\sigma L_j)$ neutrons, which caused formation of electrons $D_e(E_i N_i)$ with the discrete energies and probability of formation, probability of escape of electrons from a material of converter $H_e(Re_i\Theta_i)$ in view of the maximal free path length of electrons in a material of the converter and angle of their emission. Generally probabilities are necessary for summarizing, so calculations are made for 446 discrete energies of electrons with the energies in the range from 0.9 up to 1000 keV.

$$\Sigma E = \Sigma P_n(\sigma L_j) * \Sigma D_e (E_i N_i) * \Sigma H_e(Re_i\Theta_i) \qquad (2)$$

2.1.1. Probability of Neutrons Absorption

The attenuation of the narrow collimated neutron beam in thin layer materials is governed by the exponential law [12]

$$F_x = F_0 \exp(-N_A \sigma X) \qquad (3)$$

where F_x and F_0 are the neutron flux density after and before its passage through the layer of the material with the thickness X, correspondingly, N_A is the number of nucleus in the volume of 1 cm^3, σ- is full microscopic cross-section of neutron interaction with the nuclei of material.

One can simplify the formulae for the neutron flux density on the distance R (cm or g/cm^2) from the dot isotropic neutron source, emitting I_0, neglecting the exponential correction on attenuation when $\lambda \geq\geq R$ [13]

$$F_R = I_0 / 4\pi R^2 \qquad (4)$$

In our calculations, we use four fixed neutron energies. These are neutrons with wavelengths of 1, 1.8, 3 and 4 A^0. In Table 1 their wavelengths, corresponding energies (eV) and velocities (m/s) are shown.

Natural gadolinium is a mixture of isotopes that could participate in the (n,γ) nuclear reaction. The basic characteristics of the most widespread isotopes gadolinium including cross-section of interaction with neutrons, daughter isotopes, and a half-life period of unstable daughter isotopes, are presented in Table 2 [14,15].

As one can see in this table, the most interesting for our calculations are natural Gd and its 155 and 157 isotopes, which have abnormally high cross-sections of interaction with neutrons. Other isotopes give an insignificant contribution to the interaction with neutrons.

In Figure 1 the dependence of cross-section (barn) on the energy of incident neutrons (eV), for natural gadolinium and the same dependence for its 157 isotope are shown. Arrows indicate energy of neutrons for which we will carry out our calculations. As one can see, with the reduction of energy of neutrons, the cross-section of interaction strongly increases. Especially it increases in the region of cold and ultra cold neutrons.

Table 1.

Wavelength A°	Energy of neutrons (eV)	Velocity of neutrons (m/c)	Capture Cross-Section (barns) for NatGd	Capture Cross-Section (barns) for 157Gd
1	0,081894	3955	13 563.56	75 323.47
1,8	0,025276	2197,2	48 149.41	253 778.40
3	0,0090993	1318,3	70 597.77	367 842.60
4	0,0051184	988,76	89 066.84	464 373.40

Table 2.

Isotope	Abundance (%)	Cross-section (b)	Daughter isotope	T1/2
^{nat}Gd	100	48890	-	-
^{152}Gd	0.2	1100	^{153}Gd	241.6 d
^{154}Gd	2.2	90	^{155}Gd	Stable
^{155}Gd	14.7	61000	^{156}Gd	Stable
^{156}Gd	20.6	2.0	^{157}Gd	Stable
^{157}Gd	15.68	255000	^{158}Gd	Stable
^{158}Gd	24.9	2.4	^{159}Gd	18.6 h
^{160}Gd	21.9	0.8	^{161}Gd	3.66 min

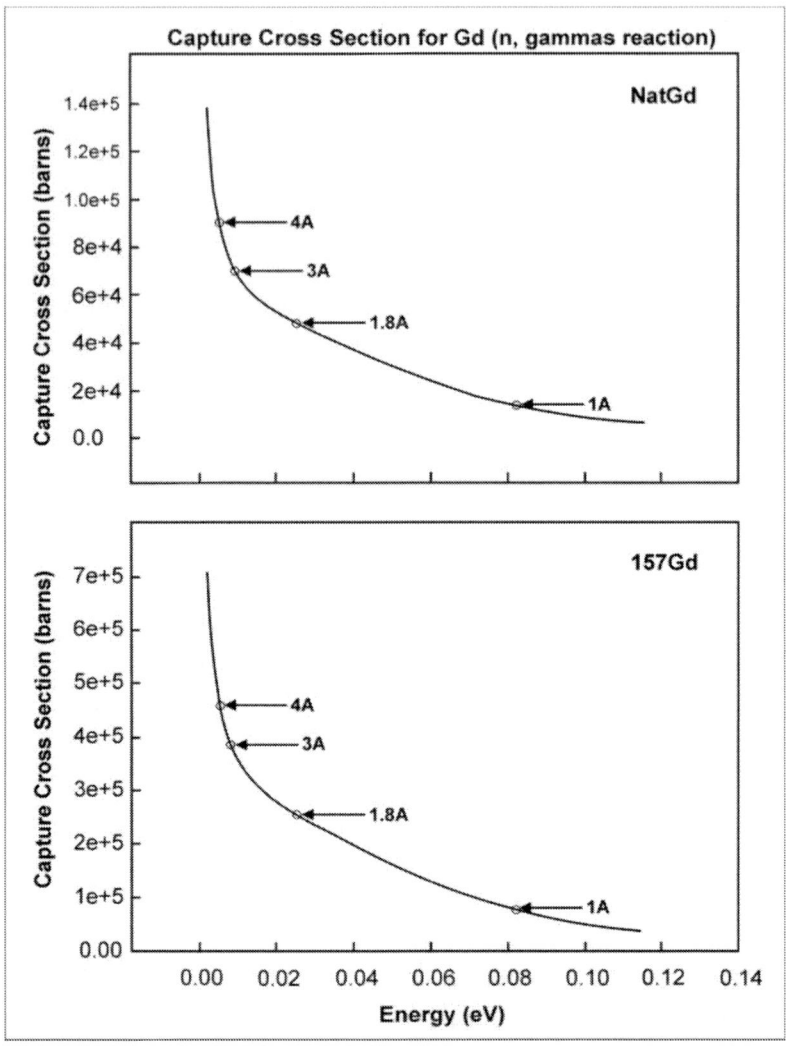

Figure 1. Cross-section of thermal neutrons capture, for the reaction (n,γ), depending on energy of neutrons for natural gadolinium and its 157 isotope [16].

2.1.2. Probability of Gamma Quanta's Formation

In the reaction of ^{157}Gd neutron capture, 7937.33 keV energy is emitted. In total 390 lines with energy ranges from 79.5 up to 7857.670 keV with line intensity of 2×10^{-8} up to 139 gamma-quanta on 100 captured neutrons are emitted. In Table 3, the most intensive, low-energy gamma-lines having high coefficient of internal conversion are presented [17]. In Figure 2, the histogram showing dependence of gamma quantum intensity on the energy is presented.

Table 3.

Isotope	Daughter isotope	Eγ [kev]	Cross-section [b] (error)	Iγ [1/100 n] (error)
157-Gd	158-Gd	79.510	4010(100)	77.3(19)
157-Gd	158-Gd	135.26	38(4)	0.73(8)
157-Gd	158-Gd	181.931	7200(300)	139(6)
157-Gd	158-Gd	212.97	10.8(7)	0.21(13)
157-Gd	158-Gd	218.225	55(4)	1.06(8)
157-Gd	158-Gd	230.23	20.0(11)	0.385(21)
157-Gd	158-Gd	255.654	350(19)	6.7(4)
157-Gd	158-Gd	277.544	493(12)	9.50(23)
157-Gd	158-Gd	365	59(5)	1.14(10)
157-Gd	158-Gd	780.14	1010(22)	19.5(4)
157-Gd	158-Gd	944.09	3090(70)	59.5(13)
157-Gd	158-Gd	960	2050(130)	39.5(25)
157-Gd	158-Gd	975	1440(21)	27.8(4)

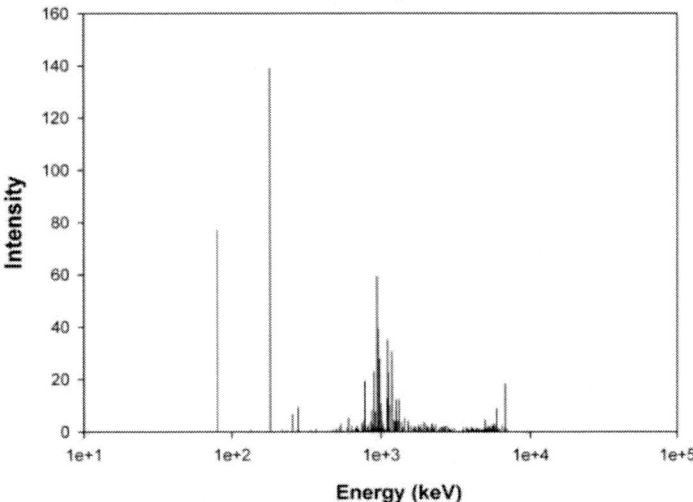

Figure 2. Histogram of dependence of intensity of gamma-quanta on energy for the reaction of ^{157}Gd (n, γ) ^{158}Gd (by 100 neutrons).

2.1.3. Probability of Internal Conversion Electrons Formation

As there are the low-energy quanta at the spectrum during their emission electrons from an atomic subshells (so-called internal conversion electrons) are radiated with a high probability. The nuclear removes its excitation by radiating a gamma-quantum, but also there a close located electron can be irradiated. Usually K-electron (electron from K-subshell) is emitted, but also electrons from the higher subshells (like L, M, N and so on) can be emitted. Vacancy of electrons (an electronic hole), formed as a result of this process, is filled by another electron from a higher level. This process is accompanied by radiation of X-ray quantum, or radiation of Auger-electron.

The probability of formation of internal conversion electrons can be considered as follows. An electromagnetic decay of the atomic nucleus can proceed by competing modes: electromagnetic radiation (γ), production of electron-positron pairs (e^+e^-) or emission of orbital electrons (e^-) that is, internal conversion. The conversion coefficient α is the ratio of the electron emission rate (T_e) to the gamma emission rate (T_γ),

$$\alpha = T_e / T_\gamma \qquad (5)$$

The values of α are depend on four parameters: (1) the charge of the decaying nucleus, (2) the energy of the nuclear transition, (3) the atomic subshell out of which the orbital electron is ejected and, finally, (4) the multi-polarity and parity of the nuclear transition. The knowledge of the coefficients is one of the most important tools for the determination of parity and multipolarity of electromagnetic nuclear transitions and the construction of nuclear decay schemes, but other applications exist as well.

The emission of nuclear gamma rays is accompanied by the emission of orbital electrons. Their branching ratio is the conversion coefficient α. The discovery of this process and its naming as a "conversion of the γ-radiation" is due to Hahn and Meitner [18]. The first correct theoretical description is due to Hulme, in 1932. A review of the theory was given recently by Pauli, Alder, and Steffen [19]. In the lowest, non-trivial order of perturbation theory, the conversion coefficient for a transition of pure electric multipole order L is given by

$$\alpha_\sigma(EL) = \pi a \omega \sum_k \frac{(2j+1)(2j+1)}{L(L+1)} \begin{pmatrix} j_0 & j & L \\ 1/2 & -1/2 & 0 \end{pmatrix}^2 \times$$
$$RkkEL + TkkOEL2 \qquad (6)$$

The index σ refers to the atomic orbital (subshell) out of which the electron is ejected. The quantities in big parentheses are 3-*j* symbols. The total conversion coefficient is given by

$$\alpha_\tau = \sum_i \alpha_{\sigma_i} \qquad (7)$$

and the total electromagnetic decay rate by

$$T = T_\gamma + T_e = T_\gamma(1 + \alpha_t) \qquad (8)$$

for transition energies ω < 2$m_e c^2$.

The relativistic angular-momentum quantum number is related to the total angular momentum *j* by $j = |k| - 1/2$. An index zero refers to the initial bound state. A subshell σ is represented in addition by n_0, the principal quantum number, i.e., σ=(n_0, k_0). The so-called dynamic radial matrix elements T_{kko} contain nuclear transition currents and charges,

$$T_{kk_0} = 0 \qquad (9)$$

For later reference, we introduce a few other quantities. The bound energy of the bound electron with energy W_0 is given by

$$\epsilon_b = 1 - W_0 \qquad (10)$$

The kinetic energy of the free electron is given by

$$E = W - 1 \qquad (11)$$

The transition energy ω is related to the latter two by

$$\Delta E = \omega = W - W_0 = E + \epsilon_b \qquad (12)$$

For a full subshell, the occupation probability has the value $\omega_\sigma = 2j + 1$. For a broken subshell, ω_σ has a smaller value caused either by a smaller number of valence electrons or by punching holes into a subshell. For the neutral atom, of course

$$\sum_\sigma \omega_\sigma = Z \qquad (13)$$

The effect of internal conversion is accompanied by the significant X-ray radiation, which could positively affect the use of scintillation detectors with the fine-dispersed gadolinium. We will only consider electrons with the energies higher than 20 keV.

Data on coefficients of internal conversion are different in various sources [20,21] that lead to the divergences in quantity of the secondary electrons. In our last modeling, we based on the last data presented in the database [22].

The most intensive lines electrons are allocated in Table 4.

Table 4.

Electron Energies (keV)	Electron output 1/100 n	Electron path in Gadolinium (μm)	Energy of primary gamma quantum	Comment, level
29.3	35,58	4,7	79.51	K
34.9	13	6,29		K-Auger
71.7	5,57	20,7	79.51	L
78	1,2	23,78	79.51	M
131.7	6,96	55,70	181.93	K
174.1	0,99	86,27	181.93	L
180.4	0,21	91,23	181.93	M
205.4	0,14	111,47	255.66	K
227.3	0,16	130,27	277.54	K
729.9	0,03	649,38	780.14	K
893.85	0,06	830,05	944.09	K
911.8	0,04	849,83	960	K
926.8	0,03	866,35	975.4	K

In Table 4, data on the most intensive lines of electrons with the probability of emission higher than 0.03/100 neutrons, for the energies of primary gamma quantum less than 1 MeV. Data on Auger electrons, which are formed during the K-shell filling, are presented as well. In our calculations, totally 446 discrete electron energies with the output probability of more than

10^{-5} on 100 incidence neutrons were considered. On Figure 3, the histogram is presented, which shows the dependence of the most intensive lines of electron intensity on their energy.

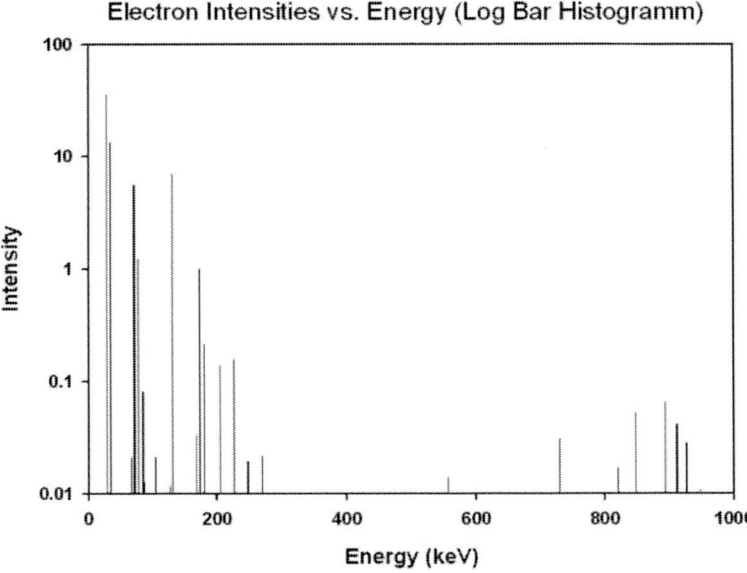

Figure 3. Intensity of internal conversion electrons emitted in the reaction 157Gd (n,γ) 158Gd depending on their energy (by 100 neutrons).

2.1.3. Intensity of Auger Electrons

Energy of emitted electrons defined by the energy of outcoming gamma quantum and the bind energy of electrons on the atom subshells

$$E_e = E_\gamma - E_{bi} \tag{14}$$

Table 5.

Z	K	L1	L2	L3	M1	M2	M3	M4	M5	N1	N2	O1	O2
$_{64}$Gd	50.24	8.38	7.93	7.24	1.89	1.69	1.55	1.22	1.19	0.38	0.27	0.04	0.03

where, E_e is the energy of outcoming electron, E_γ – energy of gamma quantum, E_{bi}- bound energy of electrons on the atom shells.

In Table 5, given bind energy of electrons on different gadolinium atom shells [15].

By knocking out electrons, an electron hole (vacancy) is formed, which is filled by electrons from higher levels. During vacancy filling X-ray radiation, with the energy equal to the difference of bind energies at corresponding levels, is radiated

$$E_x = E_{bi} - E_{bj} \qquad (15)$$

where E_x is the energy of X-ray radiation, E_{bi} and E_{bj} are bind energies of electrons at the corresponding atomic levels. Energy of excitation of atom can be removed also by the emission of Auger electrons. These electrons are emitted instead of X-ray quantum and possess energy equal to the energy of X-ray quantum minus bind energy of electron at the corresponding level

$$E_{ea} = E_x - E_{bi} \qquad (16)$$

where E_{ea}– energy of Auger electron.

In Table 6, the most intensive X-ray lines and their relative outputs [23] are represented. Data are normalized to the intensity of line Ka1, whose intensity is chosen as 100%. Also, Augur electron data are represented, irradiation of them results in reduction of X-ray radiation output.

The effect of internal conversion is accompanied by the significant X-ray radiation, which could positively affect on the use of scintillation detectors with fine-dispersed gadolinium.

Table 6.

X-ray radiation			Auger electrons		
X-ray energy (average) (keV)	IX (%)	Comment	Electron Energy (keV)	Ie (%)	Comment
6.06	42.6	L	4.84	201	L- Auger
42.31	55.6	Ka2 (KL2)	34.9	14.2	K- Auger
42.9962	100	Ka1 (KL3)			
48.7	30.8	Kb1 (KM(tot))			
50.0	8.9	Kb2 (KNO(tot))			

Probability of formation of Auger electrons have been derived from the theoretical emission probabilities through the relationship

$$I_{K-XY} = 100(1 - \omega_K)\frac{P_{K-XY}}{\sum P_{K-XY}}, \qquad (17)$$

where PK-XY is the theoretical emission probability of a K-XY Auger electron, from Chen et al. [24], ω_k is the K fluorescence yield, from Krause [25], and the summation is over all Auger electrons that are energetically possible.

Approximate Auger-electron energies can be calculated by the empirical Dillman quotation [26].

The average energy for a K-L_iX Auger transition is given by

$$\bar{E}(K - L_i X) = E_K - E_{L_i} - E_{M_3} - 0.75\left(E_{M_3+} - E_{M_3}\right) \qquad (18)$$

and for higher atomic shells by

$$\bar{E}(K - XY) = E_K - 2E_{M_3} - 0.75\left(E_{M_3+} - E_{M_3}\right). \qquad (19)$$

E_{li} is the bind energy of the L_i atomic shell for the element, E_K and E_{M3} are the corresponding bind energies of the K and M_3 atomic shells, E_{M3+} is the bind energy of the M_3 atomic subshell for the next higher element, and X and Y are designations for the higher atomic subshells. For more precise Auger electron energies, one is referred to the publication of Larkins [27].

2.1.6. Passage of Electrons through a Substance

The probability of escape of electrons from a material of the converter can be considered on the basis of Bloch theory. For realization of calculations, it is important to have exact data on brake ability of various substances for the charged particles. Brake ability is average speed of lose of energy by charged particles in any point along their tracks.

For electrons and positrons, full brake ability usually is divided on two components: a - the brake ability caused by collisions ("collision stopping power"), - average losses of energy on the unit of length of path due to non-elastic Coulomb collisions with the bound electrons of the environment, resulting in ionization and excitation; b - radiating brake ability ("radiative stopping power") - average losses of energy on the unit of length of path due to emission of brake radiation in the electric field of atomic nucleus and atomic electrons. Division of brake ability into two components is expedient for two reasons. First, methods of their determining are completely various. Second, the energy going on ionization and excitation of atoms is absorbed by the environment rather close to the track of a particle, while the basic part of energy lost in the form of brake radiation leaves far from a track before to be swallowed up. This distinction is important, when the attention is accented to the energy "transferred locally" to environment along a track, in distinction to the energy lost by an incident particle. Actually, the part of energy lost in ionization impacts, turns to kinetic energy of secondary electrons, and is transferred thus to some distance from a track of an initial particle.

In order to receive a rough estimation of locally transferred energy, it is expedient to enter the limited brake ability caused by collisions ("restricted collision stopping power") and to determine it as average losses of energy on the unit of length of the path due to events of excitation and ionization, in which the energy transmitted to secondary electrons, less than the chosen threshold.

Though brake abilities and path of electrons are widely used, they are seldom measured and turn out from the theory of energy losses. All data brought in tables contain the brake ability of electrons caused by collisions at the energies higher 10 keV, which is calculated according to Bethe theory (1930, 1932, 1933). Energy in 10 keV usually is considered as the bottom limit of applicability of the theory. Basically, the non-trivial value describing properties of the substance in Bethe formula for brake ability is mean excitation energy, representing geometrical average of energies of excitation (atoms) of the substance, weighed in view of the appropriate oscillator forces. Except for the elements with very small nuclear number mean excitation energies are approximately equal to 10Z eV, where Z is the nuclear number. Exact calculations abinitio of mean excitation energies are possible now only for simple atomic gases. For the majority of substances, the mean excitation energy could be determined from experimental data.

Other important value in the formula for the brake ability, not contained in original Bethe theory, is correction on the density effect, reducing brake ability at polarization of substance by the relativistic charged particles [28]. Also in calculations, it is necessary to take into account corrections on density effect by the Sternheimer method (1952) [29].

The linear brake ability caused by collisions with the dimension energy/length, is designated as — $(dE/dx)_{col}$ or S_{col}. It is frequently more convenient to consider the appropriate mass brake ability caused by collisions, S_{col}/ρ, where ρ is the substance density. Transition from the linear to the mass brake ability caused by collisions substantially removes dependence on the density except for residual dependence due to the correction on density effect. If S_{col} is expressed in MeV $\cdot$ cm^{-1} and ρ – in g$\cdot$cm^3 S_{col}/ρ is expressed in MeV $\cdot$ cm$^2 \cdot$ g^{-1}.

The brake ability caused by collisions is realized due to the energy transmitted by a incidence particle to connected nuclear electrons. We shall designate by $d\sigma/dW$ differential cross-section (on a atom electron) of non-elastic impacts with the energy transfer W. Then the mass brake ability caused by collisions is

$$\frac{1}{\rho} S_{co1} = NZ \int W \frac{d\sigma}{dW} dW, \tag{20}$$

where N is the number of atoms by 1 g of substance; Z is the nuclear number; $N = N_A/M_A = (uA)^{-1}$, where $N_A = 6,022045 \times 10^{23}$ mol^{-1} – Avogadro number; M_A is molar weight, g x mol^{-1}; A is the relative weight of atom (sometimes designated A_r) and $u = 1.6605655 \cdot 10^{-24}$ g is nuclear mass unit (1/12 weights of nuclide atom ^{12}C). Following formalism of Uehling (1954)[30], Bethe discuss results of estimation of the equation of brake ability [expression (20)]. These results are applicable to electrons and positrons, mesons, protons, α-particles and the heavy ions, completely deprived electrons. Energy W transferred to nuclear electrons in non-elastic impacts divides on two classes depending on those, is W less or more than some threshold W_c which should satisfy to two conditions: a) W_c should be big in comparison with bond energy of nuclear electrons of the braking substance; b) the parameters of impact appropriate to losses of energy, smaller than W_c, should be big in comparison with the sizes of atoms.

Then the mass brake ability caused by collisions could be represented as the sum two components:

$$\frac{1}{\rho} S_{co1} = \frac{1}{\rho} S_{co1}(W < W_c) + \frac{1}{\rho} S_{co1}(W > W_c). \tag{21}$$

The basic result of Bethe theory with reference to electrons and the heavy charged particles could be expressed by the formula

$$\frac{1}{\rho}S_{co1}(W < W_c) = \frac{2\pi r_c^2 mc^2}{u}\frac{1}{\beta^2} \times \frac{Z}{A}z^2\left[ln\left(\frac{2mc^2\beta^2 W_c}{(1-\beta^2)l^2}\right) - \beta^2\right], \quad (22)$$

where r_e is the classical radius of electron; mc^2 is the rest energy of electron; β is the speed of incidence particles in terms of speed of light; z is an initial charge of a particle in terms of a charge of electron and I- is average energy of excitation of atoms of the substance.

Using numerical values, Cohen and Taylor (1973) [31] for various physical constants, we receive that

$$\frac{2\pi r_c^2 mc^2}{u} = -\frac{(2\pi)\cdot(7.940775\cdot 10^{-26} cm^2)\cdot(0.5110034 MeV)}{1.6605655\cdot 10^{-24} g} =$$
$$0.153536\ MeV\cdot cm^2\cdot g^{-1} \quad (23)$$

Expression (23) is fair when the speed of incident particles is greater than speed of nuclear electrons. With reference to electrons K-shell, it means that the condition $(Z/137\beta) < 1$ is satisfied.

Component of brake ability owing to close collisions, it is estimated in approximation of free and bond electrons

$$\frac{1}{\rho}S_{co1}(W > W_c) = NZ\int_{W_c}^{W_m} W\frac{d\sigma}{dW}dW, \quad (24)$$

where $d\sigma/dW$ now is the differential cross-section of the energy transfer W in the collision with the free electron, and

$$W_m = \frac{2\tau(\tau+2)mc^2}{\left[1 + 2(\tau+1)^m \Big/ M + (m/M)^2\right]} \quad (25)$$

is the greatest possible energy transfer; τ is the attitude of kinetic energy, the incidence particles to their rest energy of rest; m/M is the ratio of weight of an electron to the weight of incidence particles.

For electrons, the big transfers of energy to nuclear electrons, which are considered as free, are characterized by cross-section (Moller (1932)) [32].

$$d\sigma = \frac{2\pi r_e^2 mc^2}{\beta^2}\frac{dW}{W^2}\left[1 + \frac{W^2}{(T-W)^2} + \frac{\tau^2}{(\tau+1)^2(W/T)^2} - \frac{2\tau+1}{(\tau+1)^2}\frac{W}{(T-W)}\right], (26)$$

where $\tau=T/mc^2$ is the ratio of kinetic energy of an incidence electron to its rest energy. In the cross-section, Moller takes into account both relativistic and spin effects, and exchange effects due to indistinguishability of incident electrons and electrons of a target. Conventionally, the brake ability caused by collisions is calculated for fastest of the electrons formed in impact. The greatest possible transfer of energy W_m equal to T according to expression (2.37), in this case should be equal to T/2. Using in expressions (20), (21) and (22) the Miller section, we receive the following formula for the mass brake ability of electrons caused by collisions (Rohrlich and Carlson, 1953; Uehling, 1954г.) [33, 30]:

$$\frac{1}{\rho}S_{co1} = \frac{2\pi r_e^2 mc^2}{u}\frac{1}{\beta^2}\frac{Z}{A}\left[\ln(T/I)^2 + \ln(1 + \tau/2 + F^-(\tau) - \delta)\right], (27)$$

where

$$F^-(\tau) = (1-\beta^2)\left[1 + \tau^2/8 - (2\tau+1)\ln 2\right]. \quad (28)$$

Half of the value in the square brackets in the expression (2.39) represents the brake number on nuclear electron $L(\beta)$.

Corrections to Bethe theory, similar to corrections on shell-effect for brake ability, were discussed by Inokuti (1971) [34] concerning cross-section of excitation. He marks that these corrections contain the additional member proportional to the ratio m/M of electron mass to the mass of incident particles. Thus, one can expect, that in the case of electrons, for which $M = m$, corrections will be much more than for protons. Probably the same happens concerning corrections of brake ability on shell-effect.

From measurements of brake ability $(S_{co1}/\rho)_{exp}$, it is possible to determine the mean excitation energy:

$$\ln I = \ln\left(\frac{2mc^2\beta^2}{1-\beta^2}\right) - \beta^2 - x - \left[\frac{(S_{co1}/\rho)_{exp}}{\frac{4\pi r_e^2 mc^2}{u}\frac{1}{\beta^2}\frac{Z}{A}Z^2}\right]. \quad (29)$$

Where

$$x = \frac{C}{Z + \frac{\delta}{2} - zL_1 - z^2 L_2} \qquad (30)$$

is the full correction uniting the corrections on shell-effect and the density effect, correction of Barkas and the Bloch correction.

Let ΔS_{col} is the error of the measured value S_{col}, and Δx is the error of a correction term x. We assume that Δx and S_{col} are independent and could be incorporated quadratialy, then the total error of estimated value I makes

$$\Delta I = I[(\Delta L_{exp})^{\uparrow}2 + (\Delta x)^{\uparrow}2]^{\uparrow}(1/2) = I[(\Delta S_{\downarrow}co1/S_{\downarrow}co1\uparrow 2 L_{\downarrow}exp\uparrow 2 + \Delta x\uparrow 2\uparrow 1 2, \qquad (31)$$

where L_{exp} is experimentally determined brake number [last term in the expression (31)], and ΔL_{exp} is the appropriate error. When in order to determine the values I one uses the data on run, the analysis of errors is more complicated and should take into account both the error of experimental value of run and the error of the correction term x at all energies down to initial energy of incidence particles.

The full correction C on the shell effect is the sum of components C_K, C_L... for various nuclear shells. It is supposed that the error owing to application of hydrogen-like wave functions is rather small for K-shell, is more essential to L-shell (especially for nuclear numbers $Z < 30$) and, probably, it is even more for the M - environment.

Extending of calculations on farther shells is possible by application of more perfect wave functions, but could be very difficult. Bichsel [35], instead of this method, applied semi-empirical numerical approach with the parameters determined from experimental data on the brake ability. He has accepted that up to factor dependences C_M on speed of a particle shell effect are similar, and this assumption is extended also to farther shells. For L-shell with eight functions, Walske [36] has received the function $C_L(\theta_L, \eta_L)$, which depends on potential of ionization of L - shell through the parameter θ_L and on the energy of particle via the value

$$\eta_L = \left(\beta/\alpha Z\right)^2, \qquad (32)$$

where $Z^* = Z - 4,15$ is effective nuclear charge for L-shell. The correction for the M – shell is calculated by the following ratio:

$$C_M = V_m C_L(\theta_L, H_M \eta_L), \tag{33}$$

where $V_M = 1/8$ of numbers of electrons on the M-shell, and H_M is adjustment parameter. The similar scale ratios were used for N-shell and for the combined O-P-shell.

In Figures 4 and 5, dependence of size of free path length electrons in gadolinium (g/sm2) and (μm) from their energy [37] is presented. Size Re can be defined by energy of electron and size specific ionization losses. The absorption coefficient F_0 characterizes probability of absorption of electrons in the substance. If X is the thickness of the converter, Re is the electron path in the material of the converter, then

$$F_0(X) = 1 - X^* \rho / Re, \tag{34}$$

where ρ is the density of the converter. For the gadolinium it is $\rho = 7,9$ g/cm^3.

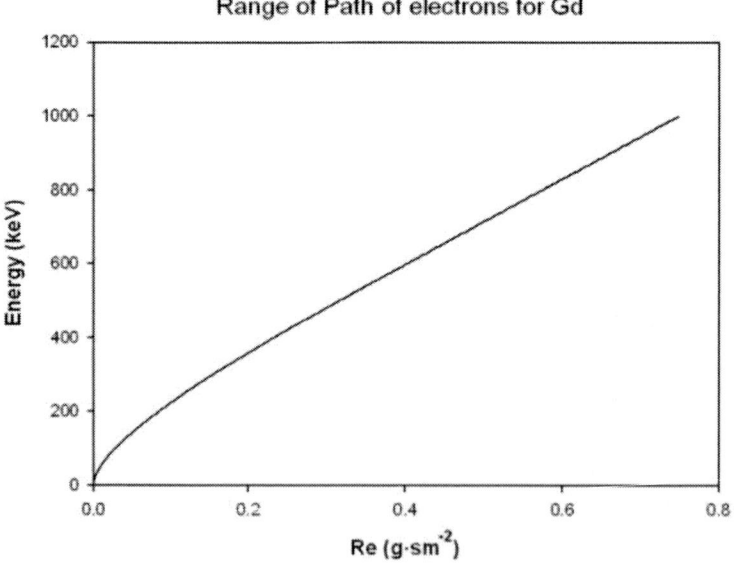

Figure 4. Dependence of the free path length of electrons on their energies in gadolinium (g/cm^2).

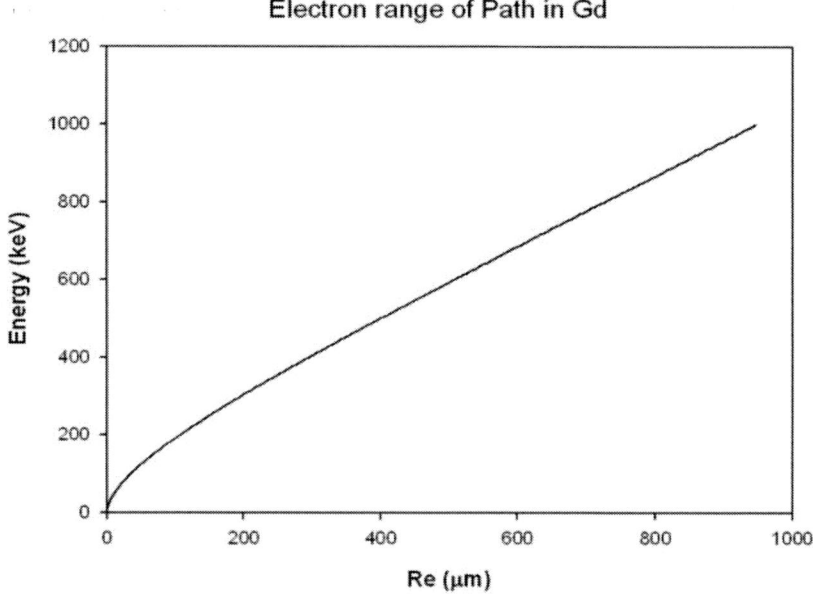

Figure 5. Dependence of the free path length of electrons on their energies in gadolinium (μm).

Re can be defined both by the electron energy and by the specific ionization loss value

$$Re = \int_0^E dE(-dE/dX) \qquad (35)$$

2.2.1. Model Representation and Calculation

Modeling was realized by examining the simplest targets, namely the plane-parallel foils. The calculations for thermal neutrons with the fixed energies, which correspond to the neutron wavelength of 1, 1.8, 3 and 4 A^0 were carried out; also, both the thickness of converters from 1 μm to 40 μm, and isotopic composition of converter (for natural Gd and ^{157}Gd) were varied. In the calculation, all electrons (appeared as a result of neutron capture act) are taken into account, which are able to escape an infinite plane-parallel plate of the converter. The ratio of the number of electrons escaped from the foil to that of the number of incident neutrons is referred to as the efficiency of the converter.

Conventionally, we divide the thickness of a foil into more thin components. For each elementary layer, we count the probability of neutrons absorption with the fixed energy. Efficiency of the converter will be determined by the sum of probabilities of neutron absorption and probability of the electron escape from the converter. In order to calculate the electron escape, we have chosen a simple model, namely, geometrical. The choice of the model is made from the following assumptions: all electron emissions are isotropic, the length of path for any fixed energy of electrons (Re_i) is constant (fluctuation of power losses in the end of path is neglected). Then the density of probability to find electrons in the material forms a sphere with the radius equal to Re_i (for each fixed energy). If the center of the sphere is crossed by the plane, two identical hemispheres are formed, which correspond to the electron escape to the forward and backward hemisphere, thus the area of hemispheres could be considered as the probability of an electron output. In this case, total probability is 100%, and escape to the one of hemispheres is 50%. If we begin to cross a sphere with a step much less than Re_i, segments will be formed whose area will be equal to probability of the electron escape. The step of iterations should be at least 100 times less than Re_i, and then the electron absorption under the big angles could be neglected. The area of a segment and accordingly probability of the electron escape becomes equal to zero in the intersection of a sphere by the plane at the distance Re_i. The sum of probabilities of electron escapes for all energies, taking into account their weight contribution, will determine the total electron escape probability.

Separated energies of electrons could be divided into four groups (Figure 4). In the first group, the least energetic and therefore having small ranges in the substance, and in the fourth one the most energetic electrons are placed.

Calculations of escape probability of isotropic emitting electrons for gadolinium are conducted. In Figure 6, one can see contributions of electrons of different energetic groups. Calculations were made for conditionally fixed point of conversion of neutrons having coordinate (X=0, Y=0). The contribution of four groups of electrons with different energies is well seen.

In calculations, the probability of neutron absorption for each elementary layer, as a result of reaction of inelastic interaction (n,γ), is determined, using database for fixed neutron energies. The probability of electron emission and their escape probability from the body of the converter were calculated.

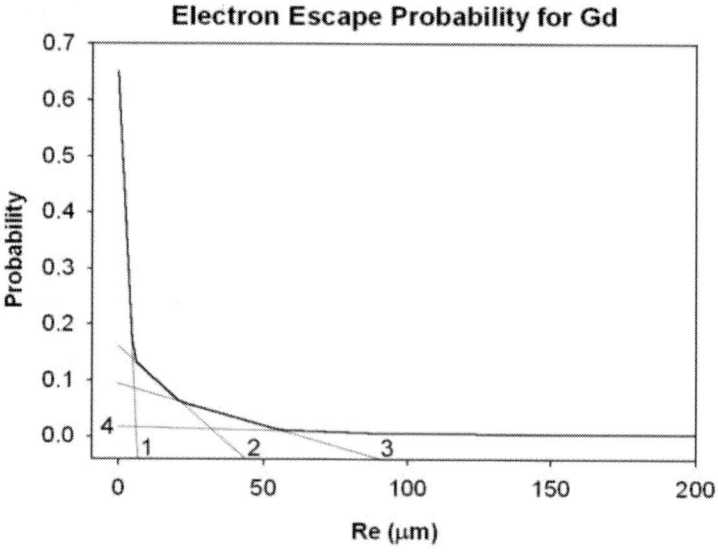

Figure 6. Probability of yield of isotropic emitting secondary electrons from gadolinium foils of various thickness, in view of their intensity and a spatial angle.

Table 7. Calculated data on efficiency and optimal thicknesses for natural gadolinium converter foils

λ(A0)	Efficiency E and optimal thickness T		(μm)
	Forward	backward	total
1	0.044 (18)	0.064 (40)	0.10 (24)
1.8	0.096 (5)	0.135 (30)	0.208 (7)
3	0.122 (4)	0.164 (25)	0.261 (5)
4	0.138 (4)	0.182 (20)	0.295 (4)

Table 8. Calculated data on efficiency and optimal thicknesses for ^{157}Gd converter foils

(A0)	Efficiency E and optimal thickness		T (μm)
	Forward	backward	total
1	0.126 (4)	0.169 (18)	0.270 (5)
1.8	0.208 (3)	0.258 (12)	0.448 (4)
3	0.232 (2)	0.277 (7)	0.489 (3)
4	0.241 (2)	0.287 (5)	0.516 (2)

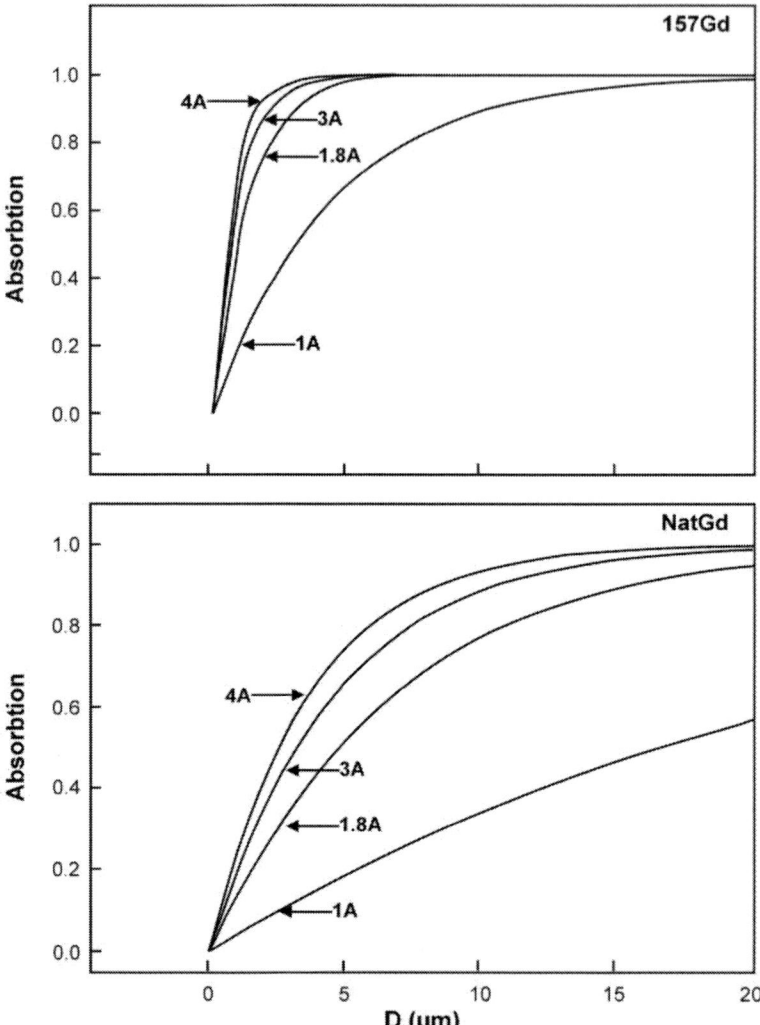

Figure 7. Curve uptakes of neutrons for lengths of waves 1, 1.8, 3 and 4 A^0 for natural gadolinium and its 157 isotopes.

Results of calculations of neutron absorption in converters made from natural gadolinium and its 157 isotope are shown in Figure 7. Figure shows that at 30 microns thickness of the natural gadolinium converter the neutrons with the wave length above 1.8 A^0 are absorbed completely. For its 157 isotope for neutrons with the wave lengths >1.8 A^0 the same absorption happens at 8 microns thickness of the converter.

In order to detect thermal neutrons by the ^{157}Gd converters, we could limit ourselves by the thickness of 5 microns of the converter, if there are no technological restrictions. It should be taken into account that the majority of electrons emitted in the reaction of radiating capture of neutrons have ranges less than 5 microns. When using natural gadolinium, the situation is more complex, since low absorption requires converter to be thicker than 20-40 micron.

Obtained data on efficiency and optimal thicknesses of converters for natural gadolinium and its 157 isotopes; see Table 7 and Table 8, correspondingly.

Figure 8 and Figure 9 show the dependence of registration efficiency on converter thickness for neutrons with the different energies.

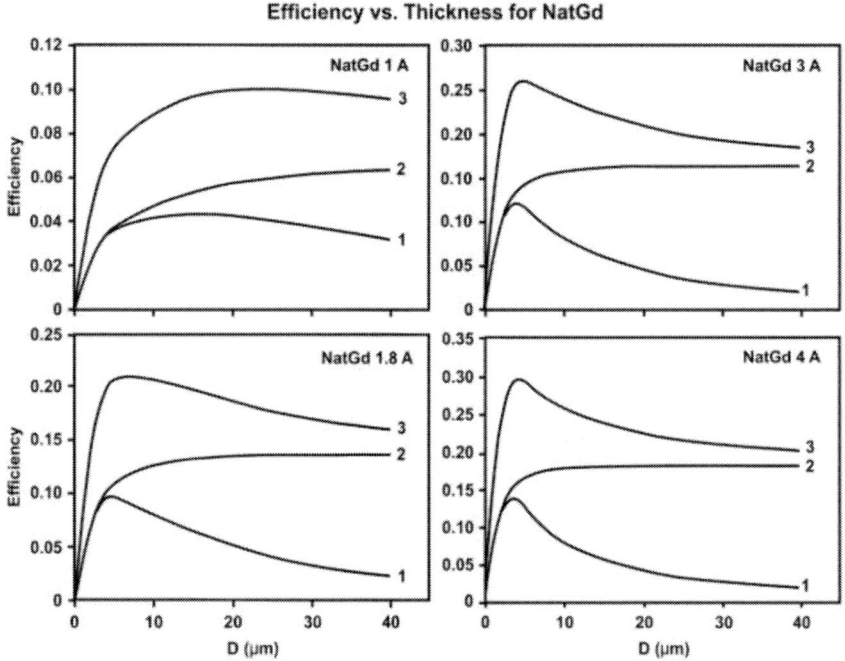

Figure 8. The dependence of the neutron registration efficiency of the natural gadolinium converter on its thickness. Curves 1 and 2 correspond to emission of electrons into the front and back hemisphere, respectively, curve 3 is their sum.

Obtained results of calculations are compared with the experimental data presented in the paper [38], which data were received in the reactors of Atominstitut in Vienna (ATI) and the ILL Grenoble.

In this paper, experimental data on the detection efficiency was measured in backward direction for six different energies and compared to a calibrated ^{3}He counter. In this work, converter made from natural gadolinium and enriched up to 90.5% ^{157}Gd converter were used.

The effect of comparison is shown in Figure 10, for natural gadolinium a curve of calculations lie a little below experimental data. Errors of calculations are caused both by the precision of determining of gamma-quanta output and by determining of neutron cross-section. For the converter made from 157 isotope, the experimental curve is little bit higher (at the account of electrons with energy more than 29 keV) that testifies to necessity of the account of more thin effects.

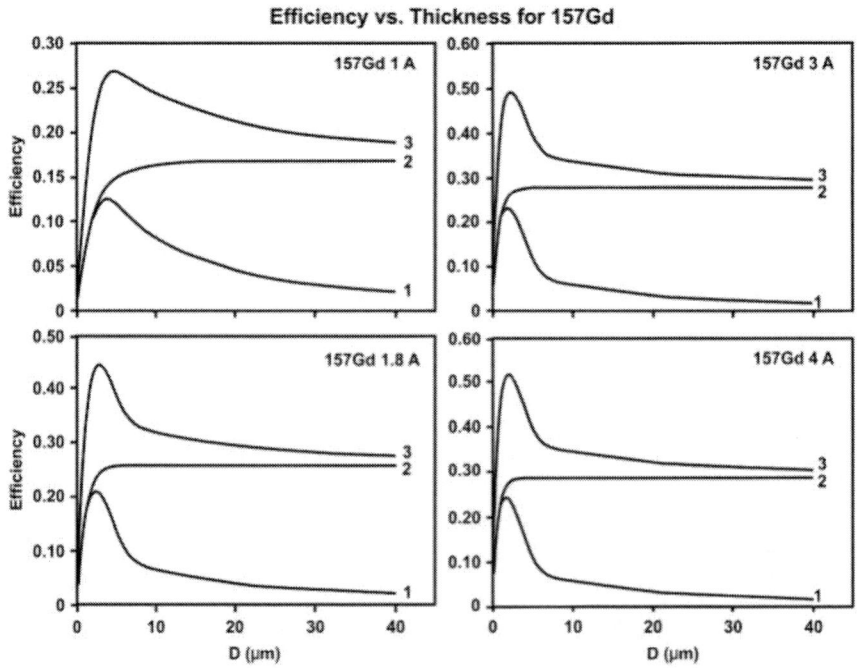

Figure 9. The dependence of the neutron registration efficiency of the 157 isotope of gadolinium converter on its thickness. Curves 1 and 2 correspond to emission of electrons into the front and back hemisphere, respectively, curve 3 is their sum.

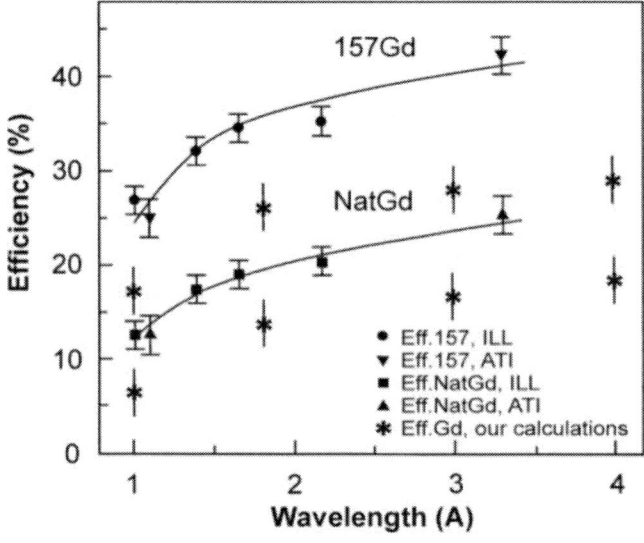

Figure 10. Comparison of our calculated data with the experimental data for backward escape geometry. Experimental data are obtained in Atominstitut in Vienna (ATI) and the ILL Grenoble [38].

2.2.2. Probability of Converters Activation

Deficiency of detectors with natural gadolinium converters is possible activation of converters, which is accompanied by radiation of retarding gamma quanta and electrons which, in turn can spot a background level. The basic contribution is carried out by long-lived isotopes (Table 1).

The activity $A(t)$ (particles s^{-1}) at a time t seconds after the removal of a thin foil from a neutron flux having an energy spectrum $\Phi(E)$ in which the foil has been irradiated for T seconds is given by [39]

$$A(t)=V\rho N_a/A(1-exp(-\lambda t)exp(-\lambda T)\int \sigma_{act}(E)\phi(E)dE \qquad (36)$$

where V, ρ and A are the volume, density and atomic weight of the foil material, respectively, N_A is Avogadro's constant, λ (s^{-1}) is the decay constant of the radioactive species produced by the irradiation and $\sigma_{act}(E)$ is the activation cross-section of the material. Values of σ_{act} averaged over the thermal and fission energy regions are given in the table, together with the activation resonance integral.

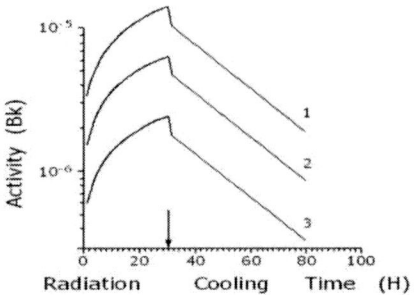

Figure 11. The background level of converters from natural gadolinium with thickness 1, 3 and 10 microns (accordingly curves 3, 2 and 1) at exposure on a beam of neutrons intensity 1n/cm^{2}s and irradiation time of 30 hours [9].

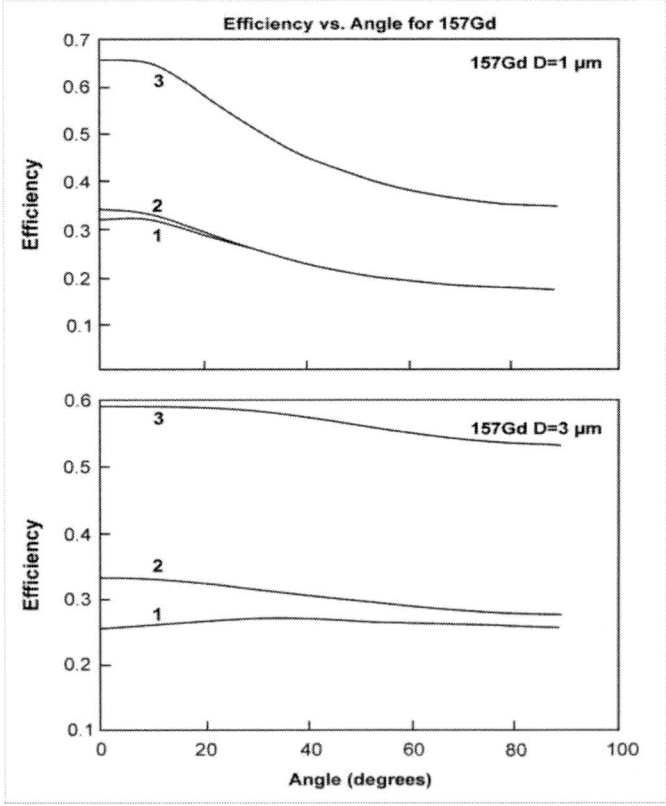

Figure 12. Dependence of efficiency of registration of neutrons with a wavelength 1.8 A from an angle of neutrons failing, for ^{157}Gd for thickness of converters 1 and 3 microns. A curves 1 and 2 accordingly an escaping of electrons in forward and back hemispheres, the curve 3 their sum.

Calculations of possible activation of converters from natural gadolinium have been fulfilled, and a background (noise level) that will accompany decay of the activated isotopes. At calculations of level of the directed background emitted electrons have been considered only. In Figure 11, effects of calculations for foils with a thickness 1, 3 and 10 microns (curves 3, 2 and 1) are presented. Calculations have been satisfied for a requirement that quantity of a neutron stream makes $1n/sm^2s$ and irradiation time of 30 hours. In its cases, maximum noise 3 microns of converters occurring as a result of activation will not exceed 4×10^{-5} Bk on an individual neutron. In time, activation level can be considered at post processing of the data.

2.2.3. Modeling for a Case of Neutrons Flow under Various Angles

In the process of calculation, we varied the angle of incidence of neutron flow onto the foil. Under the small angle of fall, we can increase neutron path in the material of the converter, yield path for the output of the secondary electrons will be small and constant. The most ideal case would be a direction of a beam of neutrons along the small thickness converter. This case cannot be realized in the practice, but yet it is possible to estimate theoretically for maximum possible efficiency define.

In Figure 12, the dependence of efficiency on the incidence angle of neutrons for ^{157}Gd for two various thicknesses of converters (1 and 3 microns) are cited. From the figure, it can be seen that at small thickness of the converter (1 micron and less) difference of electron outputs in frontward and backward of hemisphere is not so significant. The probability of an output of electrons makes approximately 32 and 35%, respectively, for an output of electrons in frontward and backward of hemispheres. Total efficiency reaches 64% [9, 40], which can be considered as the greatest possible efficiency for gadolinium converters. The difference becomes more substantial when thickness of converters is increased.

Similar calculations are made as well for converters from ^{nat}Gd. From Figure 13, it is clear that at extremely small angles of neutron incidence (2-3°), converters from natural gadolinium can compete with the converters from ^{157}Gd.

The calculations made for a case of neutron incidence under angle of 10° for various thicknesses of converters for length of a wave of neutrons 1,8°, for two types of converters natural gadolinium and its 157 isotope. Using converters from ^{157}Gd, at a thickness of 1 micron, it is possible to reach efficiency of 64% which is a theoretical limit.

Obtained as a result of modeling calculations data, allows looking with optimism at possibility of making of effective position-sensitive detectors of thermal neutrons with solid-state converters from the natural gadolinium which efficiency of registration becomes comparable with efficiency of detectors filled ^{3}He, in time, these detectors will be much more low-cost and easier to work.

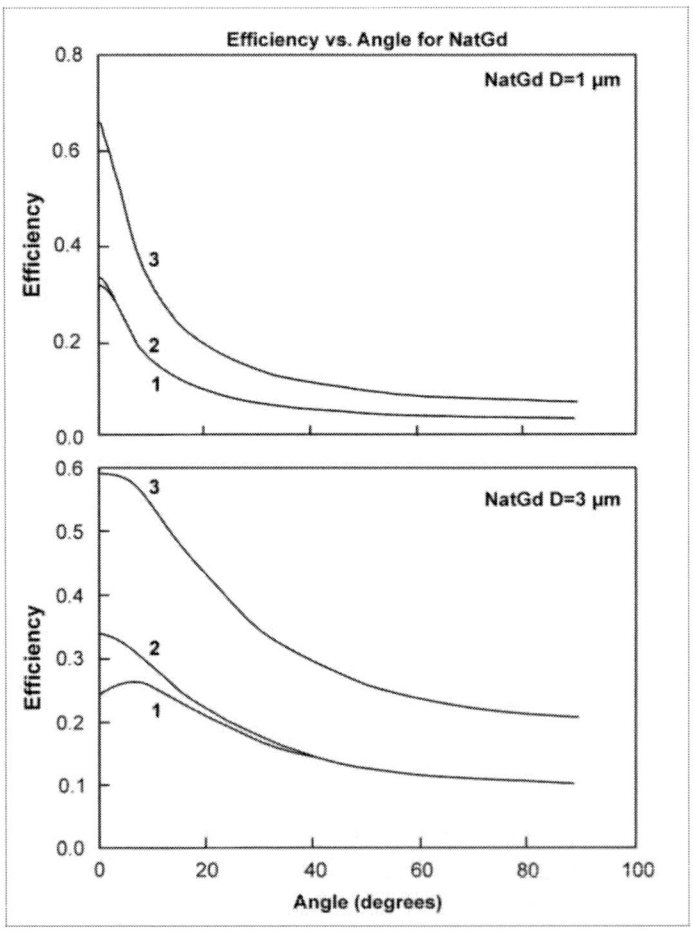

Figure 13. Dependence of efficiency of registration of neutrons with a wavelength 1.8 A^0 from an angle of neutrons failing, for ^{nat}Gd for thickness of converters 1 and 3 microns. A curves 1 and 2 accordingly an escaping of electrons in forward and back hemispheres, the curve 3 their sum.

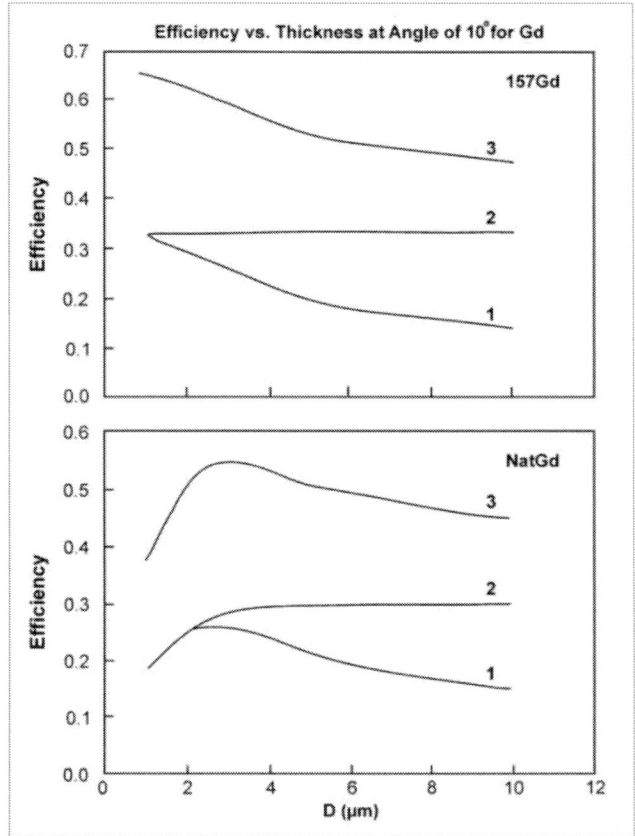

Figure 14. Dependence of efficiency of registration of thermal neutrons with a wave length 1.8 A^0 from a thickness of the converter at an angle 10^0 of flowing neutrons.

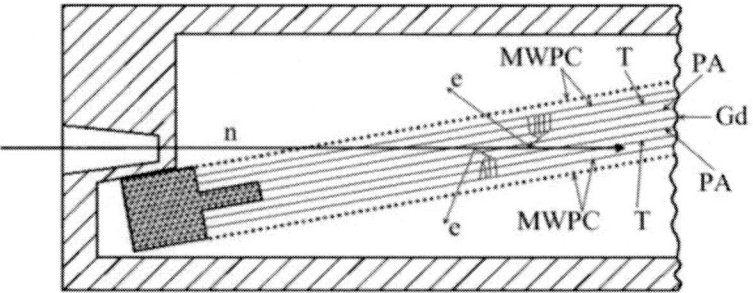

Figure 15. Schematic view of the detector, where MWPC- multiwire proportional chamber, T-transfer gap, PA- preamplification gap, Gd-gadolinium converter.

Table 9.

Sensitive area	(200-350) x 10 мм2
Efficiency of registrations	
λ=1 A	35%
λ=1.8 A	55%
λ=3 A	58%
λ=4 A	59%
Spatial resolution	0.4 мм
Time resolutions (on anode)	1-3 x10-9 c.
Gas mixture	Izo-butane
Producing pressure	5-20 torr

For example, in Figure 15, the simplified plan of the one-coordinate detector of thermal neutrons on the basis of two multistep avalanche chambers with a natural gadolinium converter is presented. The detector will be oriented at an angle 10^0 to an axis of slope of a neutron beam. In Table 9, detector key parameters [9] are given.

The detector will possess small sensitivity to gamma – irradiation and high counting rate, which will depend on a used method of removal of the information. Efficiency of registration of the detector will be comparable with detectors on a base of ^{3}He convertor at much best space resolution.

2.2.4. Contribution of Low-Energetic Electrons to General Efficiency of Converters

In our earlier calculations, we have taken into account only those electrons that have energy higher than 29 keV. These energies of electrons have been chosen not only in our calculations, but also by other authors. Thus there were not taken into account the low-energetic Auger electrons, i.e., Auger electrons from the L-shell with the energy of 4.84 keV and Auger electrons from the M-shell with the energy of 0.97 keV. These electrons have rather small free path length in gadolinium; these are 0.3 microns (4.84 keV) and 0.04 microns (0.97 keV). They bring a small contribution to the general efficiency in use of converters made from natural gadolinium as the free path length of neutrons in natural gadolinium makes tens micron. At the same time, their contribution becomes rather essential in use of converters made from 157 isotope of gadolinium as the free path length of neutrons in them does not exceed 2-3 microns and this length becomes comparable with the length of free path length of electrons.

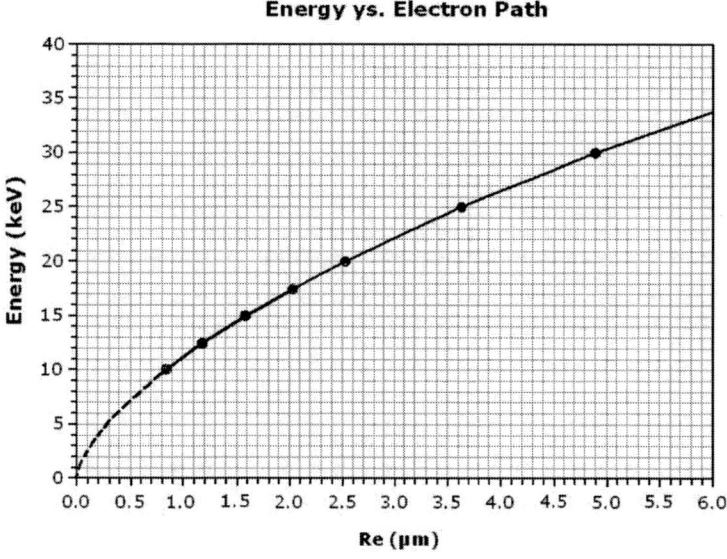

Figure 16. Curve of dependence of electron runs in gadolinium depending on their energy, for a low-energy range.

In the literature and databases, there is no data on free path length of electrons in various materials with energy less than 10 keV. All data begins with this energy. All tabular data is made on the basis of the theory Bette - Bloch and energy 10 keV is lower limit of usability of the yielded theory. For calculation of free path length of electrons of Auger in gadoliniums, we had to extrapolate available data to almost zero-point energy. Effects of extrapolation are given in Drawing 16. From the figure, it is visible that the maximum free path length of electrons with energy 4,84 keV can makes 0,3 microns, and for electrons with energy 0,97 keV can makes 0,04 microns. Similar free path length is rather small, especially in comparison with run of neutrons in natural gadolinium. In time, for converters from 157 isotopes of gadolinium run in 0,3 microns makes more than 10% from a free path length of neutrons, and it can increase efficiency of converters essentially.

There are different data on the quantity of Auger electrons in the literature, so the quantity of electrons from the K-shell makes from 10 up to 14% [14, 23]; these data are normalized to the intensity of X-ray radiation Ka1 (KL3) which intensity is chosen for 100%. The data on the quantity of electrons from the L-shell differ even more, from 150 up to 200 [14, 23]. There are no data for electrons from the M-shell at all. At the choice of quantity of electrons from the M-subshell, we started with the assumption that the quantity of Auger electrons grows at distance from a nucleus. So the quantity of electrons grows approximately 15 times at the transition from the K-subshell to L-subshell.

This tendency is kept further too, i.e., from L to M-subshell and further, N and O. At the Auger effect, the external electronic shells are peeled from the electrons practically completely. All data on the quantity of electrons are usually normalized to Ka1 (KL3) of X-ray line, which is the most intensive one. At radiating, capture of neutrons by gadolinium nuclei radiations of electrons of internal conversion occurs. The probability of emission of electrons of internal conversion makes approximately 63%. Auger electrons accompany emission of electrons of internal conversion as at radiation of electrons of internal conversion, there vacancies on electronic shells that are filled by the electrons from the higher shells are formed. This process is accompanied by X-ray radiation and Auger electrons. In our calculations, at normalization to the quantity of neutrons, at calculation of the quantity of Auger electrons, it is necessary to normalize not to Ka1 (KL3) X-ray line, but to the quantity of electrons of internal conversion, i.e., the quantity of electrons depends on factor of internal conversion.

In Table 9, the most intensive lines of electrons emitted during the process of radiating capture of thermal neutrons by gadolinium nuclei are presented. The data on Auger electrons L-Auger and M-Auger are added to the Table.

Converters made from 157 isotope of gadolinium have an abnormal high cross-section of interaction with thermal neutrons, so the cross-section makes 253778.40 barns for neutrons with the wavelength of $1.8A^0$. The cross-section strongly grows with the increase of neutron wavelength. Converters with the thickness of 2.5 microns absorb more than 80% of neutrons with the wavelength of $1.8A^0$ and more than 90% of neutrons with the wavelengths more than $3A^0$ (see Figure 7).

Another situation takes place at use of converters made from natural gadolinium. So the section of interaction makes 48149.41 barns for neutrons with the length of wave 1.8 A^0. Eighty percent of attenuation of a neutron beam ($1.8A^0$) happens at the thickness more than 12 microns.

The analysis of curves shows that at use 157 isotope of gadolinium, the small run of Auger electrons (< 0.3 microns) can increase the general efficiency of converters essentially.

Table 9.

Electron energy (keV)	Electron output 1/100 n (error)	Electron path in gadolinium- microns	Energy of initial gamma-quantum	Comments, level
0.97	>200	0.04		M-Auger
4.84	97(33)	0.3		L-Auger
29.3	35.58	4.7	79.51	K
34.9	7.9(4)	6.29		K- Auger
71.7	5.57	20.7	79.51	L
78	1.2	23.78	79.51	M
131.7	6.96	55.70	181.93	K
174.1	0.99	86.27	181.93	L
180.4	0,21	91.23	181.93	M
205.4	0,14	111,47	255.66	K
227.3	0,16	130,27	277.54	K
729.9	0,03	649,38	780.14	K
893.85	0,06	830,05	944.09	K
911.8	0,04	849,83	960	K
926.8	0,03	866,35	975.4	K

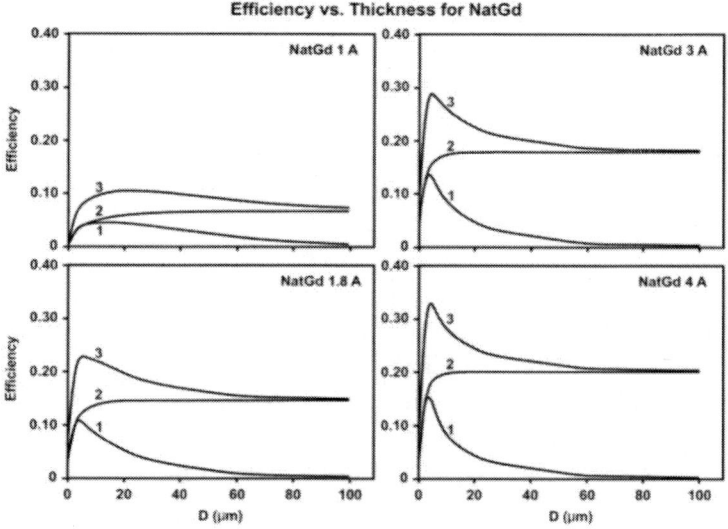

Figure 17. Curves describing the efficiency of converters made from natural gadolinium depending on the thickness of the converter, taking into account the low-energy Auger electrons. The curve 1 characterizes electron emission in a direct, the curve 2 - in back hemispheres. The curve 3 is their sum.

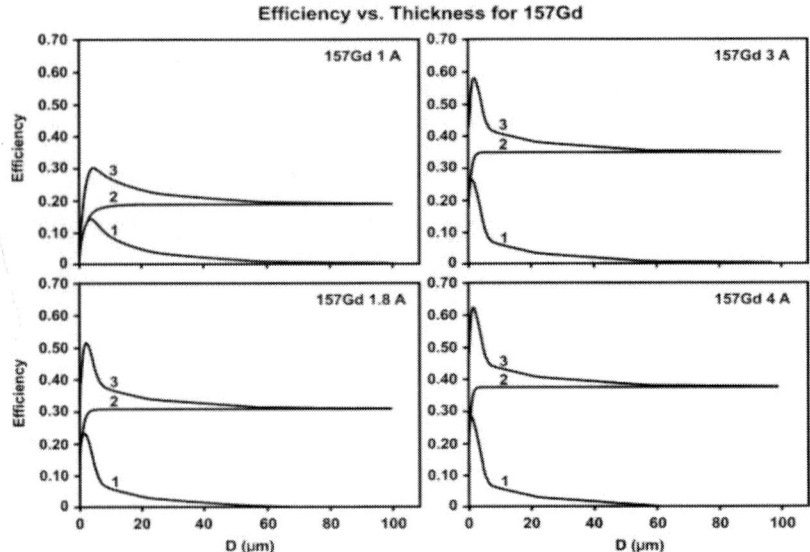

Figure 18. Curves describing the efficiency of converters made from 157 isotope of gadolinium depending on the thickness of the converter, taking into account the low-energy Auger electrons. The curve 1 characterizes electron emission in a lobby, a curve 2 - in back hemispheres. The curve 3 is their sum.

Calculations of the efficiency of gadolinium foils are carried out, at use of natural gadolinium and its 157 isotope, for four fixed wavelengths of neutrons depending on the thickness of converters. Figures 13 and 14 show the results of calculations without taking into account of the influence of low-energy Auger electrons. Figures 17 and 18 show the results of similar calculations, but here, the low-energy electrons are taken into account.

One can see from the figures that with the increase of neutron wavelength, the difference in the efficiencies is increased. Especially the difference is well visible at use of 157 isotope of gadolinium. So for neutrons with the wavelength of $4A^0$ the total efficiency grows practically by 10%. Comparison of result of calculations for a case of recording of electrons with energy is more 29 keV and more than 0,93 keV are given in Table 10.

The received data were compared with the experimental data given in the paper[38]. In this work, the experimental data on the efficiency of detecting of neutrons emitting in a back hemisphere for six various energies and their comparisons with the calibrated ^{3}He counter are received. In this work, converters made from natural gadolinium and enriched up to 90,5% ^{157}Gd were used. The work was carried out on the reactors of Atominstitut in Vienna (ATI) and the ILL Grenoble (Figure 19).

One can see from the figure that in the case of the account of the contribution of low-energy Auger electrons (electrons with the energy > 0.93 keV), the results of our calculations well coincide with the experimental data. This concurrence is well visible for the converters made from 157 isotope of gadolinium. If we do not take into account the low-energy electrons, the curve lays much below experimental data.

Table 10.

Wave length (A)	Forward	Backward	Total	Wave length (A)	Forward	Backward	Total
	^{Nat}Gd (E> 29 keV)				^{157}Gd (E> 29 keV)		
1	0,0435	0,0635	0,107	1	0,1262	0,1686	0,2948
1,8	0,0967	0,1357	0,2324	1,8	0,2098	0,258	0,4678
3	0,1218	0,1636	0,2854	3	0,2319	0,2772	0,5091
4	0,1379	0,1816	0,3195	4	0,2443	0,2865	0,5308
	^{Nat}Gd (E> 0,93 keV)				^{157}Gd (E> 0,93 keV)		
1	0,0462	0,068	0,107	1	0,1402	0,1851	0,3253
1,8	0,1063	0,1464	0,2527	1,8	0,2342	0,3103	0,5445
3	0,1335	0,1792	0,3127	3	0,2633	0,3501	0,6134
4	0,1509	0,2011	0,352	4	0,2848	0,3756	0,63

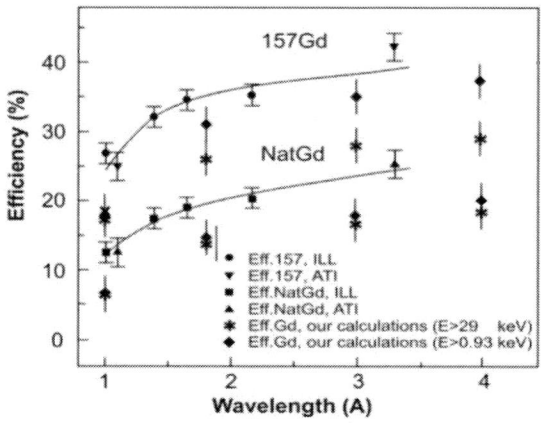

Figure 19 Comparison of the results of our calculations with the experimental data presented in the paper [38]. Calculations are conducted for two boundary energies of taken into account electrons; higher than 29 keV and higher than 0,93 keV.

2.2.5. Contribution X-Rays and Soft Gamma Radiations on General Efficiency of Converters

At calculations of the efficiency of converters, it is necessary to take into account all secondary electrons. Earlier, we took into account only internal conversion electrons and Auger electrons. But in the process, also secondary electrons appearing during absorption of X-ray radiation and gamma-quanta can participate. These are photoelectrons, Compton electrons, electron-positron pairs. During radiation of electrons of internal conversion, vacancies on electronic shells which are filled by electrons from higher shells are formed. At their filling, X-ray quanta or Auger electrons are radiated.

We have estimated the contribution of X-ray and low-energy gamma-quanta absorbed directly in the material of the converter and resulting in occurrence of secondary electrons. At practical calculations, absorption of quanta in materials of detectors also should be taken into account. At such calculation, total efficiency can appear higher, as the part of electrons can be formed from quanta in a material of the detector. In the calculations, we take into account only an escape of electrons from a material of the converter.

Internal conversion coefficient is defined as follows:

$$\alpha_{ik} = \frac{I_e}{I_\gamma},\qquad(37)$$

where I_e is the intensity of conversion electron, and I_γ is the intensity of gamma radiation. I_γ is well known (Table 3).

Energy of emitted electrons defined by energy of outcoming gamma quantum and bind energy of electrons on the atom subshells

$$E_e = E_\gamma - E_{bi},\qquad(38)$$

where, E_e is the energy of outcoming electron, E_γ – energy of gamma quantum, E_{bi} – bind energy of electrons on the atom shells.

By knocking out electrons, an electron hole (vacancy) is formed, which is filled by electrons from higher levels. During vacancy filling X-ray radiation, with the energy equal to the difference of bind energies at corresponding levels, is radiated

$$E_x = E_{bi} - E_{bj}, \tag{39}$$

where E_x is the energy of X-ray radiation, E_{bi} and E_{bj} are bind energies of electrons at the corresponding atomic levels.

Energy of excitation of atom can be removed, also due to the emission of Auger electrons. These electrons are emitted instead of X-ray quantum and possess energy equal to the energy of X-ray quantum minus bound energy of electron at the corresponding level.

Table 11.

X-Ray Energy (keV)	Shell Comments	Intensity (1/100 Vacancies)	Intensity [1/100 neutrons]
181.931	Gamma		139(6)
79.510	Gamma		77.3(19)
42,996	Kα1	47,5	28,5
42,309	Kα2	26,6	15,96
41,864	Kα3	0,00824	0,004944
48,695	Kβ1	9,3	5,58
49,959	Kβ2	3,11	1,866
48,551	Kβ3	4,81	2,886
50,099	Kβ4	0,038	0,0228
49,038	Kβ5	0,146	0,0876
50,219	KO2,3	0,45	0,27
6,058	Lα1	6,3	3,78
6,026	Lα2	0,7	0,42
6,713	Lβ1	3,9	2,34
7,102	Lβ2,15	1,32	0,792
6,832	Lβ3	0,126	0,0756
6,687	Lβ4	0,077	0,0462
7,243	Lβ5	0,0108	0,00648
6,867	Lβ6	0,063	0,0378
7,79	Lγ1	0,67	0,402
8,087	Lγ2	0,024	0,0144
8,105	Lγ3	0,034	0,0204
7,93	Lγ6	0,0055	0,0033
6,049	Lη	0,092	0,0552
5,362	Ll	0,266	0,1596

$$E_{ea} = E_x - E_{bi} \tag{40}$$

where E_{ea} – energy of Auger electron.

In the table are given both the most intensive X-ray and gamma-lines their relative escape on 100 formed vacancies [41]. We have made calculation of X-ray quanta on 100 captured neutrons (the data are presented in the table also).

The output of X-ray quanta on 100 formed vacancies is represented in Figure 20. From the figure, one can see that the most intensive are K-lines 42.3 and 43 keV. The most interesting are low-energy L-lines; however, their output is strongly suppressed by irradiated Auger electrons.

During the passage of X-ray quanta through a material of the converter, there absorption occurs. The degree of their absorption is described by the mass factor μ/ρ of attenuation of X-rays.

A narrow beam of monoenergetic photons with an incident intensity I_o, penetrating a layer of material with mass thickness x and density ρ, emerges with intensity I given by the exponential attenuation law

$$I/I_0 = exp[-(\mu/\rho)x] . \tag{41}$$

Equation (41) can be rewritten as

$$\mu/\rho = x^{-1} ln(I_0/I) \tag{42}$$

from which μ/ρ can be obtained from measured values of I_o, I and x.

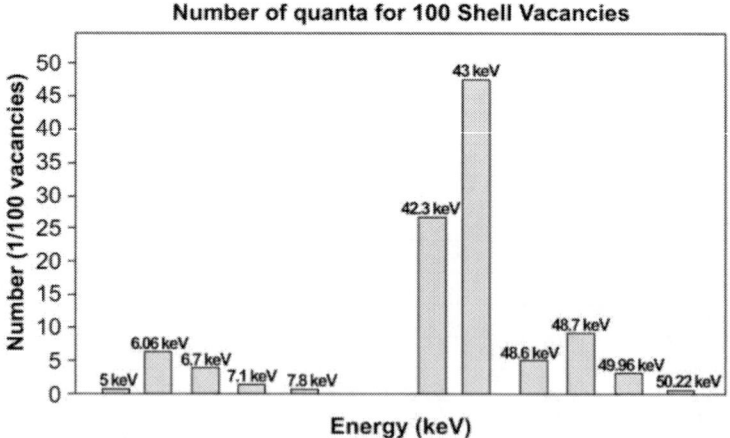

Figure 20. Amount of X-ray photons depending on their energy, which formed, on 100 vacancies in electron subshell of Gadolinium.

Note that the mass thickness is defined as the mass per unit area and is obtained by multiplying the thickness t by the density ρ, i.e., $x = \rho_t$.

Present tabulations of μ/ρ rely heavily on theoretical values for the total cross-section per atom, σ_{tot}, which is related to μ/ρ according to

$$\mu/\rho = \sigma_{tot}/uA . \qquad (43)$$

In (eq 3), u (= 1.660 540 2 × 10^{-24} g) [43] is the atomic mass unit (1/12 of the mass of an atom of the nuclide ^{12}C), A is the relative atomic mass of the target element, and σ_{tot} is the total cross-section for an interaction by the photon, frequently given in units of b/atom (barns/atom), where b = 10^{-24} cm^2.

The attenuation coefficient, photon interaction cross-sections and related quantities are functions of the photon energy. Explicit indication of this functional dependence has been omitted to improve readability.

The total cross-section can written as the sum over contributions from the principal photon interactions,

$$\sigma_{tot} = \sigma_{pe} + \sigma_{coh} + \sigma_{incoh} + \sigma_{pair} + \sigma_{trip} + \sigma_{ph.n.} , \qquad (44)$$

where σ_{pe} is the atomic photoeffect cross-section, σ_{coh} and σ_{incoh} are the coherent (Rayleigh) and the incoherent (Compton) scattering cross-sections, respectively, σ_{pair} and σ_{trip} are the cross-sections for electron-positron production in the fields of the nucleus and of the atomic electrons, respectively, and $\sigma_{ph.n.}$ is the photonuclear cross-section.

Accordingly, μ/ρ can be represented as following

$$\mu/\rho = (\sigma_{pe} + \sigma_{coh} + \sigma_{incoh} + \sigma_{pair} + \sigma_{trip})/uA . \qquad (45)$$

Values of the mass attenuation coefficient, μ/ρ, for the mixtures and compounds (assumed homogeneous) were obtained according to simple additive:

$$\mu/\rho = \Sigma_i w_i (\mu/\rho)_i , \qquad (46)$$

where w_i is the fraction by weight of the ith atomic constituent.

The methods used to calculate the mass energy-absorption coefficient, μ_{en}/ρ, are described perhaps more clearly through the use of an intermediate quantity, the mass energy-transfer coefficient, μ_{tr}/ρ.

Mass factors of absorption of X-ray quanta for various materials are represented in the tabulated data [43]. In Figure 21 the dependence of μ/ρ on energy of quanta is depicted. In the figure, peaks and edges of absorption of electronic subshells K, L, M are well visible.

Absorption of quanta strongly depends on their energy. Low-energy quanta are absorbed most intensively, however, formed secondary electrons are also the low-energy ones and have rather small run in a material of the converter.

In the below figures, dependences of formation of secondary electrons on depth of penetration of quanta are given. Amounts of formed electrons are given without taking into account their absorption in a material of the converter (an integral one).

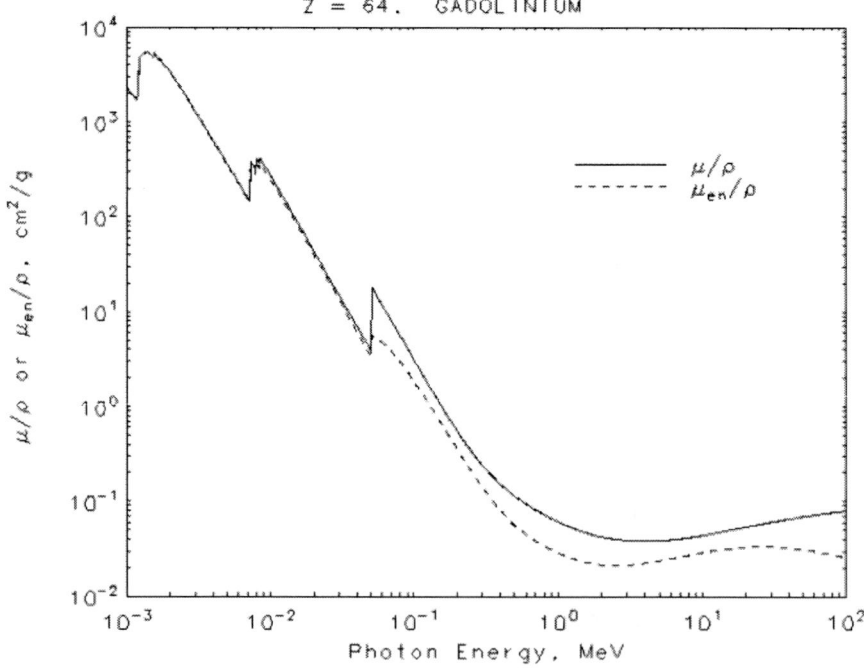

Figure 21. Dependence of mass factor of absorption of quanta on their energy for gadolinium [43].

Mathematical Modeling of Converter Performances

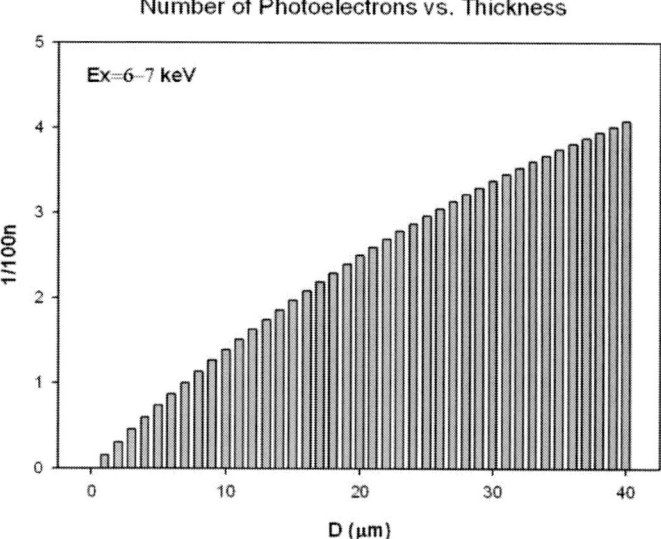

Figure 22. Dependence of number of formed electrons on thickness of the converter. For quanta with the energy range 6 up to 7 keV.

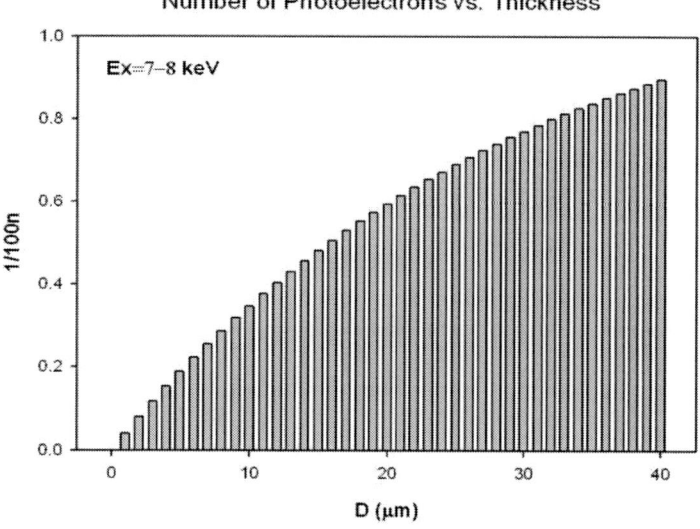

Figure 23. Dependence of number of formed electrons on thickness of the converter. For quanta with the energy range from 7 up to 8 keV.

In Figure 24, the integrated characteristic of formation of secondary electrons is presented. The low-energy quanta form mainly photoelectrons, and gamma-quantum with the energy 181.9 forms Compton electron. The low energy quanta with the energy range from 5 and up to 9 keV form low-energy electrons, the maximal free length path of which in gadolinium are accordingly equal to 0.4 and 0.6 microns for quanta with the energy 5 and 8 keV. At calculation of electron escape, the quanta formed in a layer, which is a little thicker, than a maximal free length path of electrons for the given energy, are taken into account only. In the table, the new data on electrons that are added by the data on secondary electrons formed as a result of X-ray and low-energy quanta absorption. These electrons are designated in the comment as X.

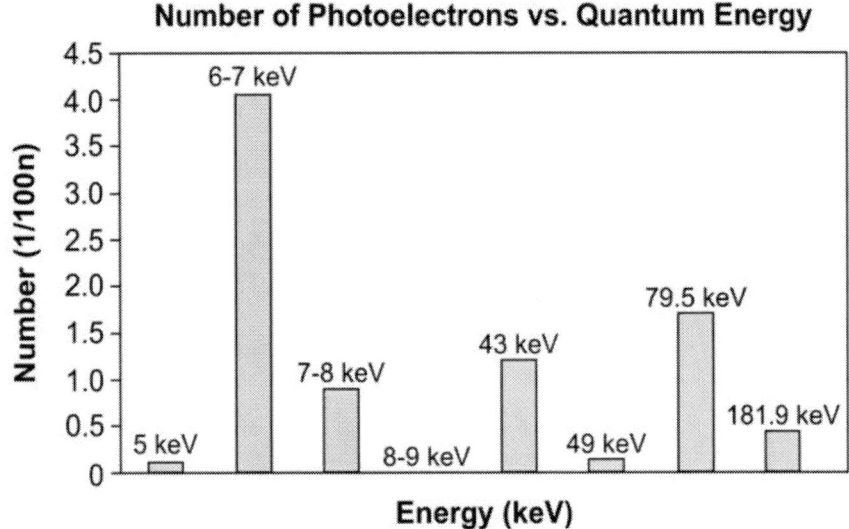

Figure 24. Dependence of the number of secondary electrons depending on energy of quanta.

The contribution of photoelectrons to a total sum of electrons for each value of energy does not exceed 2%. And their total amount does not exceed 0.97 electrons on 100 incident neutrons that is 1% less.

Calculations of the contribution of photoelectrons in general efficiency of gadolinium converters are carried out. Results of calculations give in Figure 25.

From the figure, one can see that in the case of the account of the contribution of electrons formed by X-ray quanta, the efficiency is increased a little, but their contribution is insignificant. The increase occurs for no more than by 1%. It is rather possible, that X-ray quanta can essentially affect on the total efficiency at the account of their absorption in a material of detectors. So gas detectors with Argon filling can increase efficiency essentially. If the conversion of quanta will occur in a working body of the detector, therefore the full gathering of formed secondary electrons will take place. They can render even greater influences in the case of use of semi-conductor detectors.

Table 12.

Energy of electrons (keV)	Electron escape 1/100 n (error)	Electron path in gadolinium (µm)	Energy of initial gamma-quantum	Remarks, level
0.97	>200	0,04		M-Auger
4.84	97(33)	0.3		L-Auger
5 - 8	0,2	0,4 - 0,6		X
29.3	35,58	4,7	79.51	K
29,3	0,18	4,7		X
34.9	7.9(4)	6,29		K- Auger
34,9	0,18	6,29		X
71.7	5,57	20,7	79.51	L
78	1,2	23,78	79.51	M
78	0,03	23,78		X
131.7	6,96	55,70	181.93	K
131,7	0,38	55,70		X
174.1	0,99	86,27	181.93	L
180.4	0,21	91,23	181.93	M
205.4	0,14	111,47	255.66	K
227.3	0,16	130,27	277.54	K
729.9	0,03	649,38	780.14	K
893.85	0,06	830,05	944.09	K
911.8	0,04	849,83	960	K
926.8	0,03	866,35	975.4	K

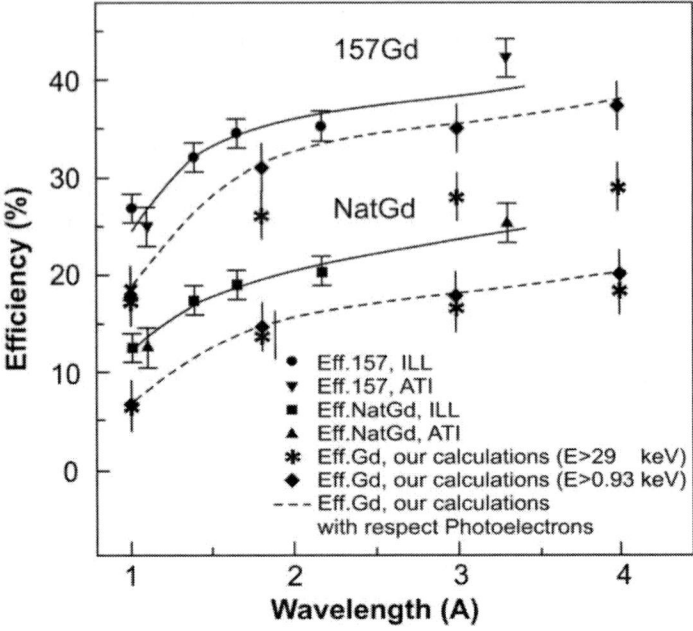

Figure 25. Comparisons of our theoretical calculations with the experimental data given in the paper [38]. Calculations are carried out for two boundary energies taken into account electrons; more than 29 keV and more than 0.93 keV and in view of the contribution of secondary electrons formed by X-ray and low-energy gamma-quanta (a dashed line).

2.3.1. Modeling of Converters Representing Sandwiches, from Supporting Films and Converters

Recently more often started to apply widely solid-state converters of thermal neutrons of complicated forms, such as bilateral converters made from thin gadolinium with supporting film made from kapton, neutron converters on the base of GEM-structures and others, which manufacturing techniques consist of drawing of thin gadolinium converters (0.5-5 microns) on the surfaces of film made from kapton with the thickness of 5-100 microns. Such converters allow manufacturing detectors of the big areas. Presence of supporting films results in additional reduction of secondary electron escapes; at the same time, the use of bilateral converters allows to increase conversion of neutrons, at use of two detectors from both sides of the converter. Modeling of efficiency of registration of thermal neutrons by these kinds of converters is an interesting problem, and during its development one can expect occurrence of new converters.

Modeling of the efficiency of registration of thermal neutrons by the converters, which represent kapton film, serving for support of thin gadolinium layers is carried out. Layers can be settled down on one, or on both sides of supporting film. Calculations are made for natural gadolinium and its 157 isotope, for four fixed energies of neutrons. In calculations, electrons with energy more than 29 keV were considered only.

Attenuation of electron flux arising in the reaction of radiating capture of neutrons by gadolinium nucleus occurs both in the substance of the converter and in kapton film. In Figure 26, curves of probabilities of electron flux output for gadolinium and kapton are given. These curves characterize absorption of electrons in these substances.

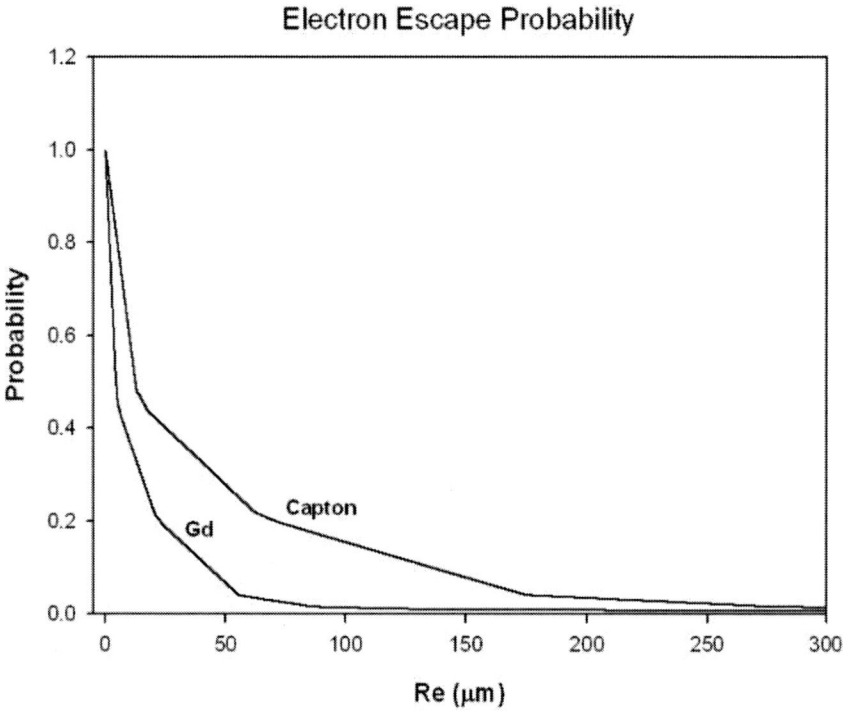

Figure 26. Probability of electron output, formed in the reaction of radiating capture of thermal neutrons by gadolinium nuclei, from kapton and gadolinium.

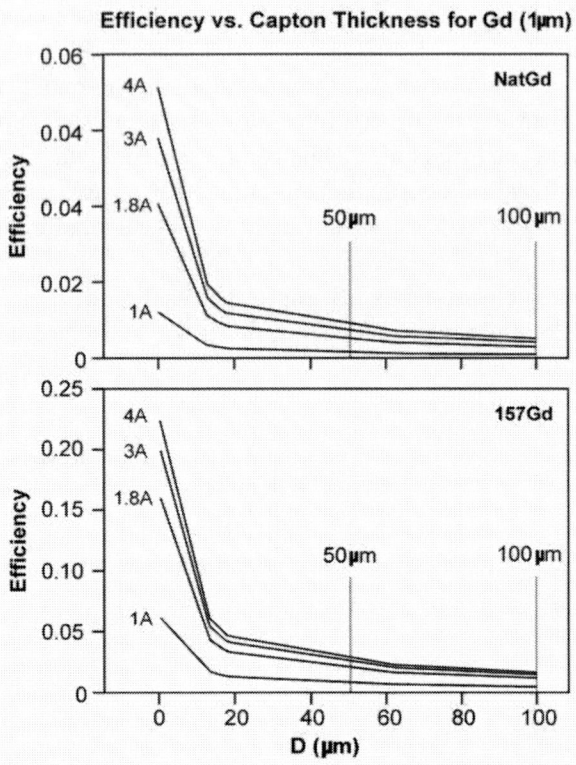

Figure 27. Efficiency of complex converters made from thin gadolinium (1 micron) put on a substrate from kapton, depending on thickness of a substrate. Various curves characterize wavelengths of neutrons.

Contributions of various energies of electrons that are well visible it is conditionally possible to part on four groups. In first group, the least energetic and having small run in substance in the fourth most energetic electrons, Figure 3.

Calculations of efficiency of converters with the thickness of 1 micron made from natural gadolinium and its 157 isotope, for four fixed neutron wavelengths are carried out. Calculations were performed for the case when only one thin converter was directly settled on the films made from kapton of various thicknesses, Figure 27. This kind of geometry is characteristic for a case of drawing of thin converting films (0.5-3 microns) on supporting substrates made from kapton. Thickness of supporting films as usual is chosen within the limits of 5-100 microns.

Model calculations of converters like the developed and used in DETNI project in Hahn-Meitner-Institut (HMI) are carried out. The detector and the converter are described in the paper [44].

The detector's size is 285 x 285 мм^2, and it is shown in Figure 28. Each segment consists of centrally located converter consisting of ^{157}Gd/CsI foil. Converters are located from both sides of supporting foil. Thicknesses of converters lay in the range from 0.5 - up to 1.5 microns of ^{157}Gd. CsI is used for thermalization of electron energy as an emitter of secondary electrons, and has thickness less than 1 micron. Necessity of the use of secondary emitter CsI caused by the reason that there plane-parallel chambers of low pressure are used in the detector. Electrons escaping from the converter have enough big range in low pressure gases, which can result in appreciable deterioration of the spatial resolution.

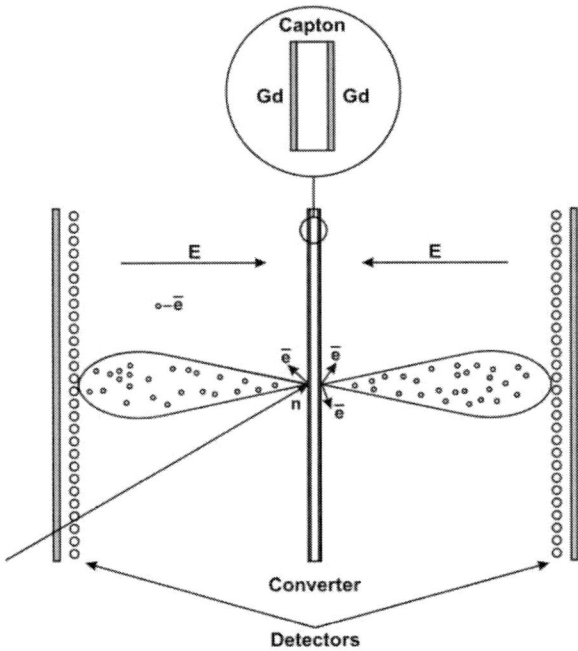

Figure 28. The scheme of the detector used in article []. The converter consists of a supporting film executed of kapton film from which two sides of its two gadolinium converter were placed.

Calculations of efficiency of registration of neutrons by gadolinium foils in this kind of geometry are carried out. Calculations were carried out for converters made from natural gadolinium and its 157 isotope, for four fixed neutron wavelengths. Thickness of converters varied from 1 up to 3 microns. Possible efficiency of converters with emission of electrons in 4π geometry without taking into account attenuation in kapton was taken into account Calculations are made for electrons with energy more than 29 KeV. At calculations, the secondary emitters of electrons executed on the basis of crystals CsI were not considered.

In Figure 31, effects of calculations of efficiency of gadolinium foil in simple geometry are given. Calculations were made for one foil without supporting films. Efficiency for converters from natural gadolinium grows with enlargement of a thickness. At use, converters from 157 isotopes of gadolinium efficiency for electrons departs in a back hemisphere not significant growth. For electrons departing for a forward hemisphere, even efficiency decrease is observed

For this complicated converter, at an arrangement of the converters from both sides of supporting film, the part of neutrons will be converted in the first converter and a part in the second. From the first converter, electrons emitting to the back hemisphere will be registered; they will be registered by the forward chamber. From the second converter, electrons emitting to the forward hemisphere will be registered, and they will be registered correspondingly by the back chamber. These processes will be the primary. At the same time, a part of high energy electrons can penetrate through the supporting film and will be registered by the next chamber. This process can increase total efficiency of registration a little.

One can see from Figures 30-31 that the optimal thickness for electron escape to a back hemisphere is the thickness of 2 microns for gadolinium-157 films. Up to 3 microns of thickness growth of efficiency is observed; however this converter is a lobby one and the further increase of its thickness will result in shielding the second converter. For the second converter (escape of electrons to the forward hemisphere), an optimum thickness is 2 microns, as the further increase of thickness does not result in increase of efficiency.

The figure shows that at use of 157 isotope of gadolinium, the main absorption of neutrons occurs on the forward converter. The second converter is substantially shielded by the first converter.

Similar calculations conducted also for converters made from natural gadolinium; results of these calculations are presented in Figures 32 and 33.

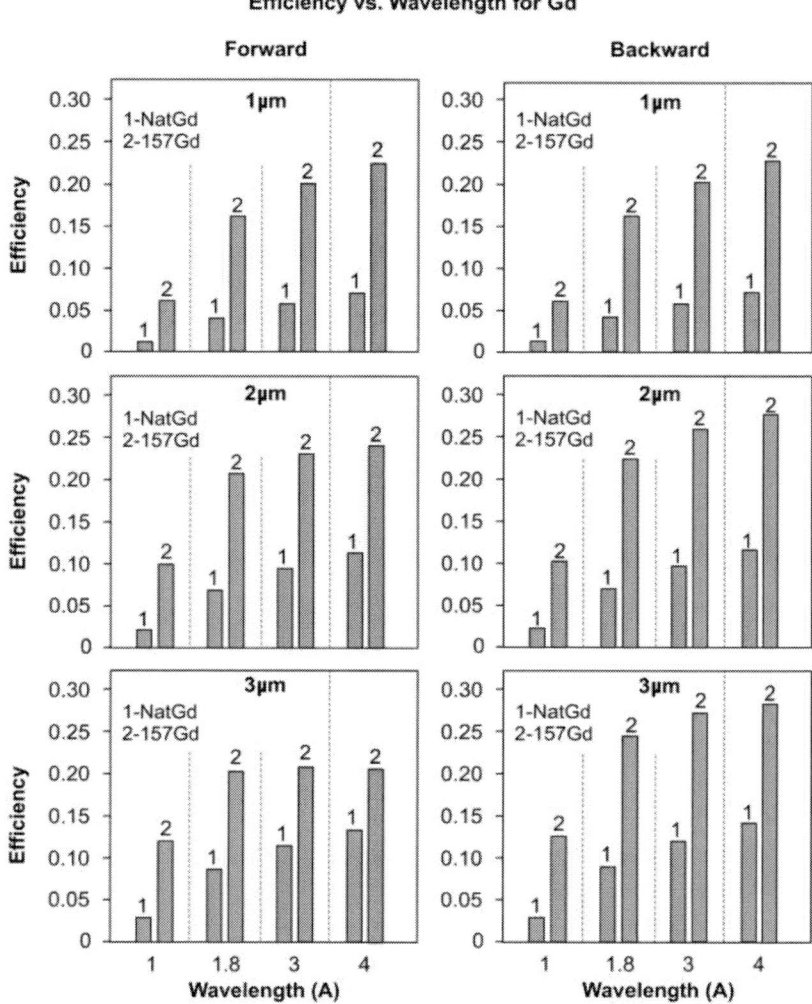

Figure 29. Efficiency of registration of neutrons with wavelengths 1, 1.8, 3 and 4 A^0 for ^{nat}Gd and ^{157}Gd for converters with the thickness 1, 2 and 3 microns without supporting films.

Calculation of efficiency thin gadolinium converters is made in view of their attenuation in kapton. Thus thickness of the kapton film (from 10 up to 50 microns) and thickness of gadolinium (from 1 up to 3 microns) were varied. From Figures 30 and 31, one can see that the optimal thickness for an electron emission to the back hemisphere is gadolinium 157 thickness of 2 microns. Up to the thickness of 3 microns growth of efficiency is observed, however this converter is a lobby and the further increase of its thickness will result in shielding the second converter. For the second converter (electron emission to the forward hemisphere), an optimum thickness is 2 microns as the further increase of thickness does not result in increase of efficiency. The optimal thickness for a kapton film is the thickness of 10 microns. With the growth of thickness, the secondary electron escape falls. Especially it has an effect at registration electrons emitted to the forward hemisphere.

In Table 13, contributions of each of two converters in the efficiency of the back (2 detector) detector, an electron emission in the forward hemisphere are shown. For a case of use of converters made from 157 gadolinium isotope. Neutron wave length is 1.8 A^0.

In Table 14, contributions of each of two converters in the efficiency of the forward detector (one detector), an electron emission to the back hemisphere are presented. For a case of use of converters made from 157 gadolinium isotope. Neutron wave length is 1.8 A^0.

In Table 15, contributions of each of two converters in the efficiency of the back detector, an electron emission to the forward hemisphere are presented. For a case of use of converters made from natural gadolinium. Neutron wave length is 1.8 A^0.

In Table 16, contributions of each of two converters in the efficiency of the forward detector, an electron emission to the back hemisphere are given. For a case of use of converters made from natural gadolinium. Neutron wave length is 1.8 A^0.

Calculations of efficiency of these kinds of converters are carried out in case of their cascading. Converters will be located the one above the other, thus each subsequent converter will be shielded by the previous converters. The case of use of three converters is considered. This kind of scheme is not applied in practice, since does not allow transferring electrons through converters to the detector. However, the case is interesting for calculations because it allows estimating a degree of shielding of the subsequent converters.

In Figures 34 and 36, the contribution of each converter for a thickness of the kapton film in 10, 20 and 50 microns is shown. One can see from these figures that at use of gadolinium-157, the basic conversion of neutrons will occur on the first converter. The subsequent converters will be substantially shielded. In Figures 35 and 37, total efficiency of similar converters is shown.

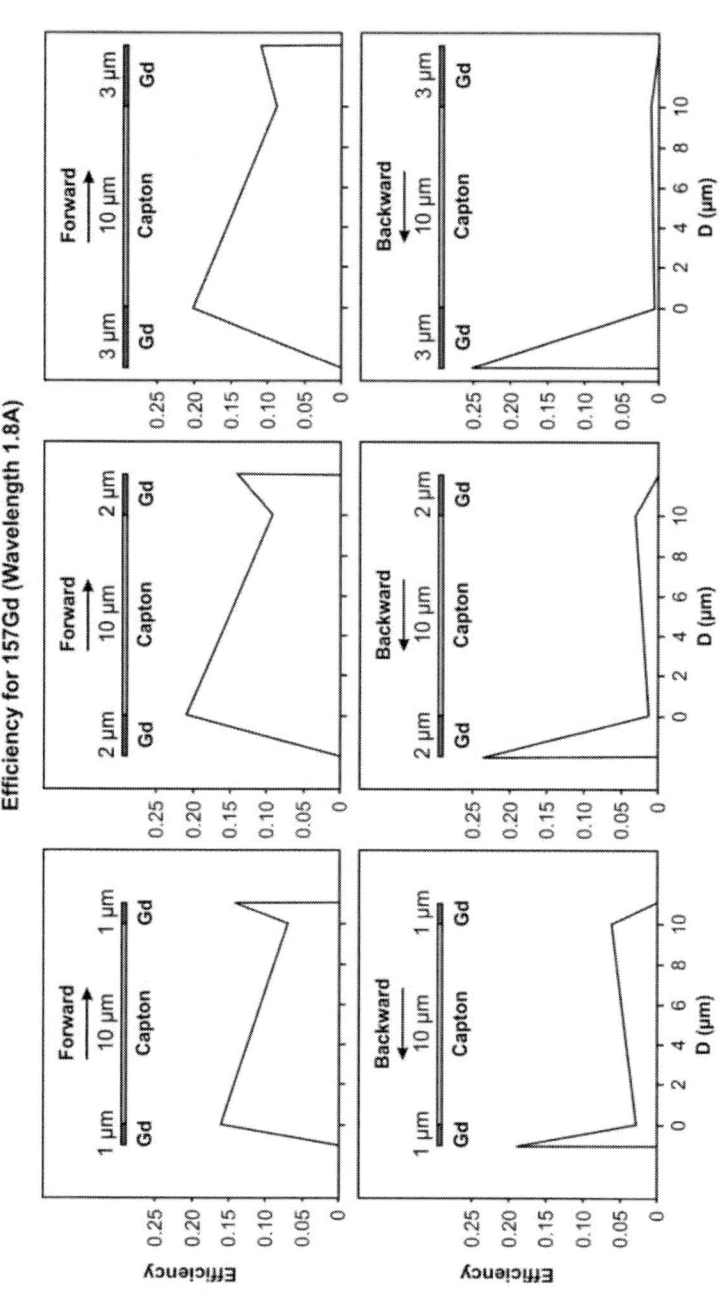

Figure 30. Efficiency of the complex converter for a case of electron output to the forward and back hemispheres is given. Thickness of the converter from ^{157}Gd is 1, 2 and 3 microns, thickness of kapton is 10 microns. Wavelength of neutrons is 1.8 A^0.

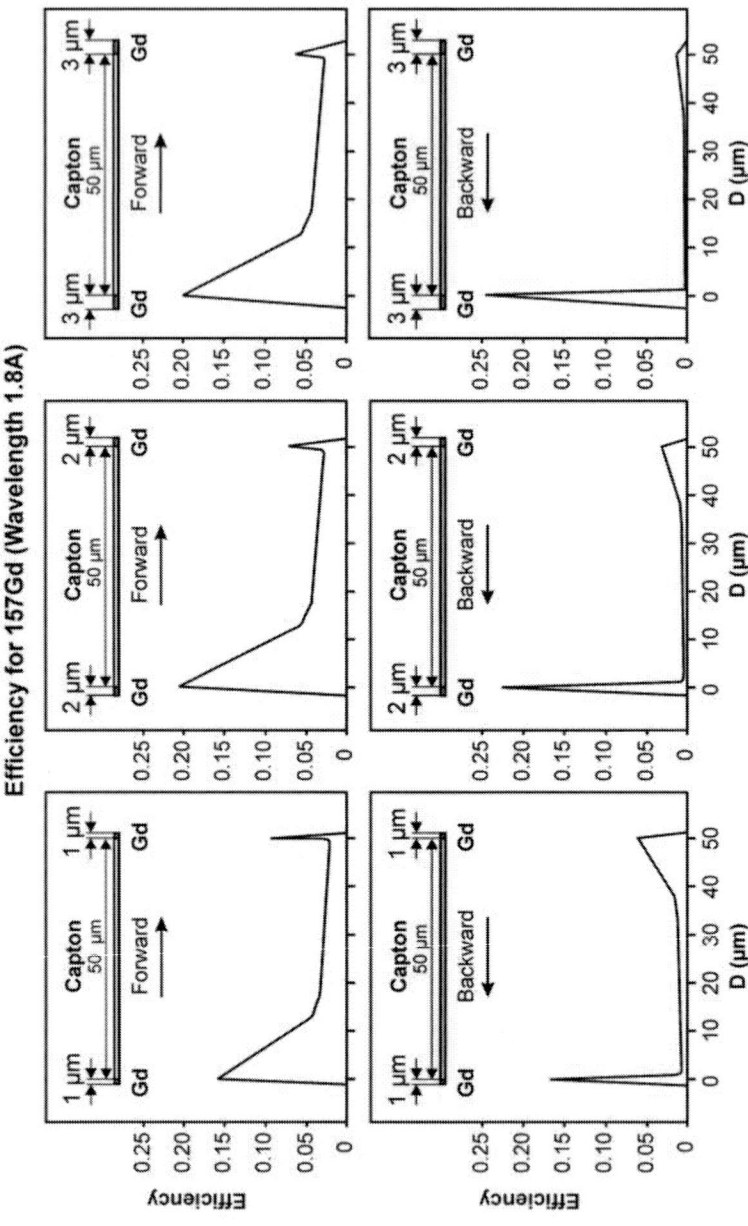

Figure 31. Efficiency of the complex converter for a case of electron output to the forward and back hemispheres is given. Thickness of the converter from ^{157}Gd is 1, 2 and 3 microns, thickness of kapton is 50 microns. Wavelength of neutrons is 1.8 A^0.

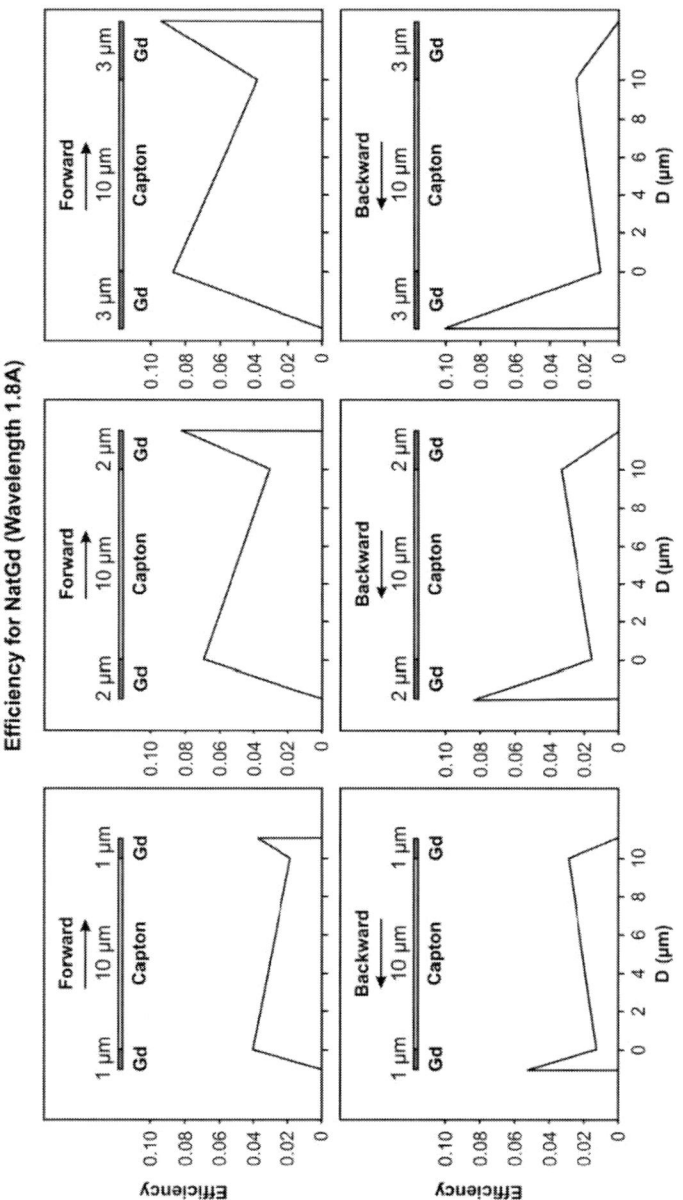

Figure 32. Efficiency of the complex converter for a case of electron output to the forward and back hemispheres is given. Thickness of the converter from ^{nat}Gd is 1, 2 and 3 microns, thickness kapton is 10 microns. Neutron wavelength is 1.8 Å^0.

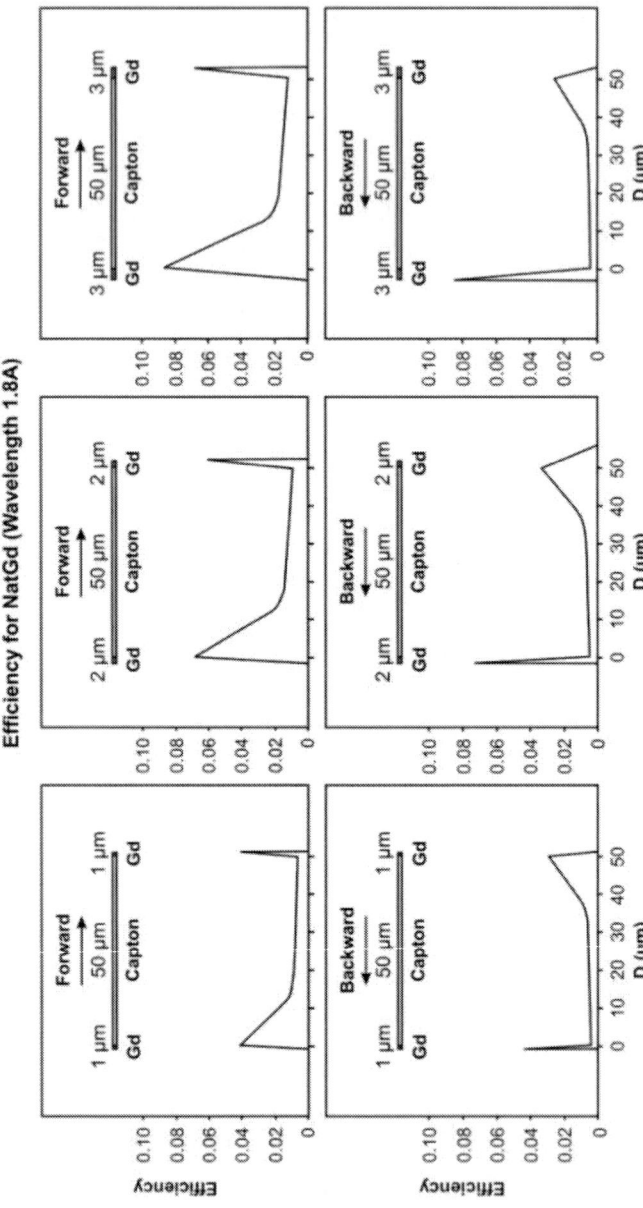

Figure 33. Efficiency of the complex converter for a case of electron output to the forward and back hemispheres is given. Thickness of the converter from ^{nat}Gd is 1, 2 and 3 microns, thickness kapton is 50 microns. Neutron wavelength is 1.8 Å^0.

Table 13.

Kapton Thickness, microns	Geometry, microns	Contribution of converters in the efficiency (forward electron emission) 2 detector			Total of both converters
		1 layer	Escape from the 1 layer in view of attenuation	2 layer	Total of both converters
10	1 + 1	0.160	0.070	0.074	0.144
	2 + 2	0.208	0.091	0.045	0.136
	3 + 3	0.202	0.089	0.020	0.109
20	1 + 1	0.160	0.032	0.074	0.106
	2 + 2	0.208	0.042	0.045	0.087
	3 + 3	0.202	0.041	0.020	0.061
30	1 + 1	0.160	0.028	0.074	0.102
	2 + 2	0.208	0.037	0.045	0.082
	3 + 3	0.202	0.036	0.020	0.056
40	1 + 1	0.160	0.025	0.074	0.099
	2 + 2	0.208	0.032	0.045	0.077
	3 + 3	0.202	0.031	0.020	0.051
50	1 + 1	0.160	0.021	0.074	0.095
	2 + 2	0.208	0.027	0.045	0.072
	3 + 3	0.202	0.026	0.020	0.046

Table 14.

Kapton Thickness, microns	Geometry, microns	Contribution of converters in the efficiency (back electron emission) 1 detector			Total of both converters
		1 layer	2 layer	Escape from the 2 layer in view of attenuation	Total of both converters
10	1 + 1	0.161	0.062	0.027	0.188
	2 +2	0.223	0.031	0.014	0.237
	3 + 3	0.246	0.011	0.005	0.251
20	1 + 1	0.161	0.062	0.012	0.173
	2 +2	0.223	0.031	0.006	0.229
	3 + 3	0.246	0.011	0.002	0.248
30	1 + 1	0.161	0.062	0.011	0.172
	2 +2	0.223	0.031	0.005	0.228
	3 + 3	0.246	0.011	0.002	0.248
40	1 + 1	0.161	0.062	0.010	0.171
	2 +2	0.223	0.031	0.005	0.228
	3 + 3	0.246	0.011	0.002	0.248
50	1 + 1	0.161	0.062	0.008	0.169
	2 +2	0.223	0.031	0.004	0.227
	3 + 3	0.246	0.011	0.001	0.247

Table 15.

Kapton Thickness, microns	Geometry, microns	Contribution of converters in the efficiency (forward electron emission) 2 detector			Total of both converters
		1 layer	Escape from the 1 layer in view of attenuation	2 layer	Total of both converters
10	1 + 1	0.041	0.018	0.035	0.053
	2 +2	0.069	0.030	0.052	0.082
	3 + 3	0.087	0.038	0.056	0.094
20	1 + 1	0.041	0.008	0.035	0.043
	2 +2	0.069	0.014	0.052	0.066
	3 + 3	0.087	0.017	0.056	0.073
30	1 + 1	0.041	0.007	0.035	0.042
	2 +2	0.069	0.012	0.052	0.064
	3 + 3	0.087	0.015	0.056	0.071
40	1 + 1	0.041	0.006	0.035	0.041
	2 +2	0.069	0.011	0.052	0.063
	3 + 3	0.087	0.013	0.056	0.069
50	1 + 1	0.041	0.005	0.035	0.04
	2 +2	0.069	0.009	0.052	0.061
	3 + 3	0.087	0.011	0.056	0.067

Table 16.

Kapton Thickness, microns	Geometry, microns	Contribution of converters in the efficiency (back electron emission) 1 detector			Total of both converters
		1 layer	Escape from the 2 layer in view of attenuation	2 layer	Total of both converters
10	1 + 1	0.040	0.013	0.029	0.053
	2 +2	0.069	0.014	0.033	0.083
	3 + 3	0.089	0.011	0.025	0.1
20	1 + 1	0.040	0.006	0.029	0.046
	2 +2	0.069	0.007	0.033	0.076
	3 + 3	0.089	0.005	0.025	0.094
30	1 + 1	0.040	0.005	0.029	0.045
	2 +2	0.069	0.006	0.033	0.075
	3 + 3	0.089	0.004	0.025	0.093
40	1 + 1	0.040	0.004	0.029	0.044
	2 +2	0.069	0.005	0.033	0.074
	3 + 3	0.089	0.004	0.025	0.093
50	1 + 1	0.040	0.004	0.029	0.044
	2 +2	0.069	0.004	0.033	0.073
	3 + 3	0.089	0.003	0.025	0.092

2.2.6. Modeling of Converters Executed from a Set Thin Drilling Converters

The new solid-state converter of thermal neutrons is offered. The converter will consist of a set of thin gadolinium foils located one over other in a gas volume. Foils there will be drilled with the fine step (2 mm) with diameter of apertures 1 mm. These foils will have an optical transparency of 40%; correspondingly, gadolinium will fill 60% of a surface. Secondary electrons will emit for all sides, and in an electric field, they will be entice up in holes and further drift in the direction of the detector. As a detector, there are various gas detectors, such as the multi-wire proportional chamber, multi-step avalanche and multi-strip detectors, and so on, can be used. The schematic sketch of this kind of detector is shown in Figure 38. In Figure 39, the cross-section of the detector is shown.

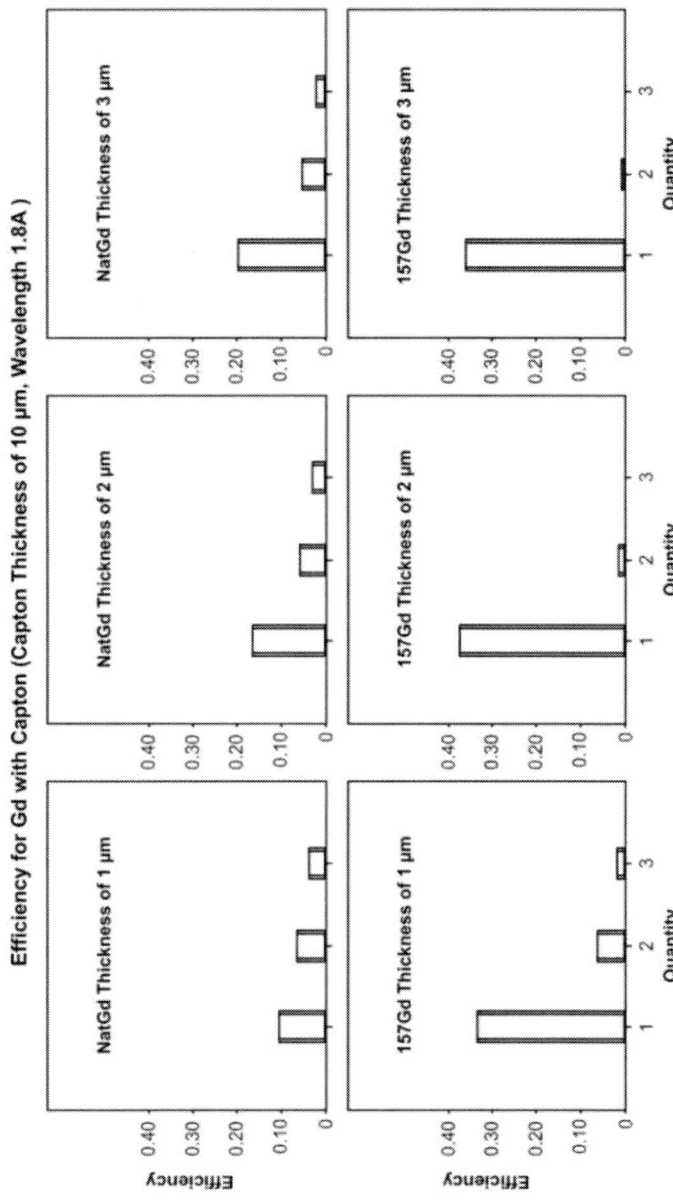

Figure 34. Efficiencies of each of the complex converters at their sequence cascading. For an application case of kapton thickness 10 micron.

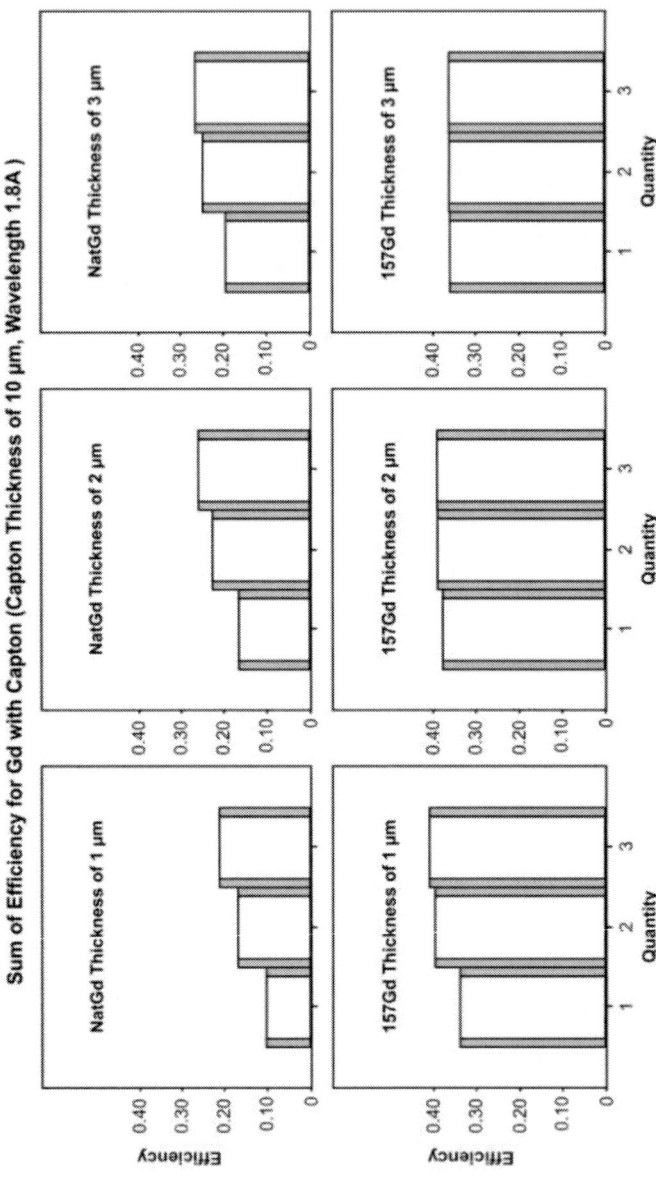

Figure 35. Summary efficiency of the complex converter at their sequence cascading. For an application case kapton a thickness 10 micron.

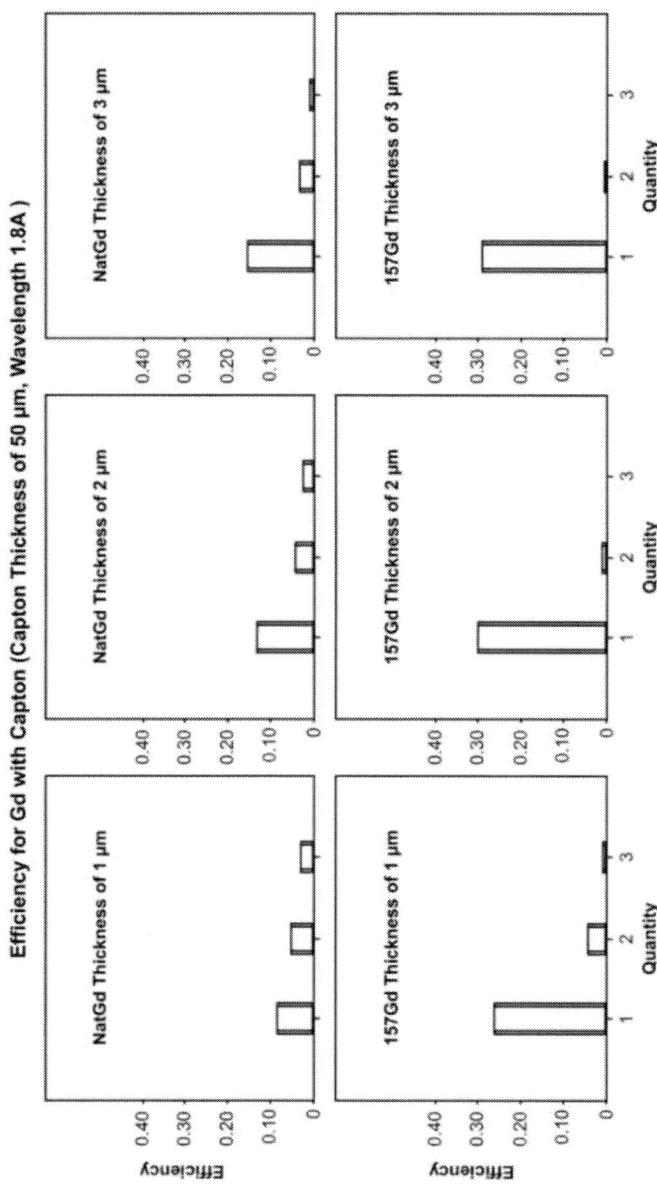

Figure 36. Efficiencies of each of the complete converters at their sequence cascading. For an application case of kapton thickness 50 micron.

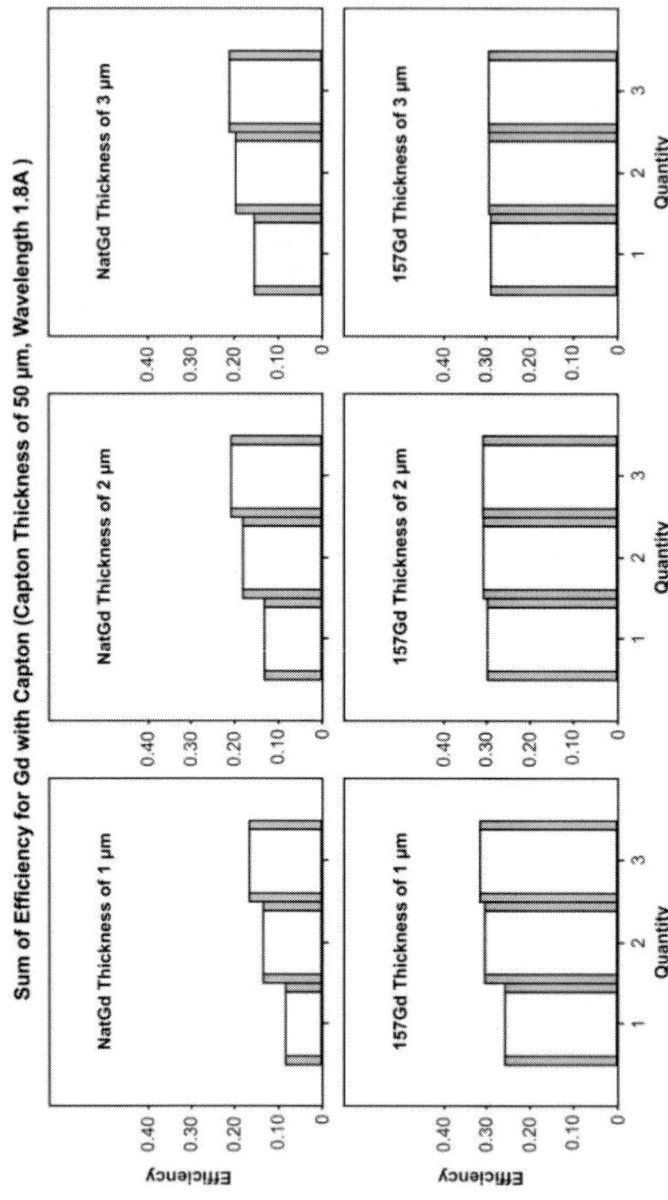

Figure 37. Summary efficiency of the complex converter at their sequence cascading. For an application case kapton a thickness 50 micron.

Mathematical Modeling of Converter Performances 67

Technologically, this kind of converter can made from foils with the thickness more than 5 microns. However, more thin foils are also of interest for calculations. The distance between the foils can be chosen in a range from 0.2 up to 2 mm.

For avalanche detectors working at normal atmospheric pressure, the distance of 0,2 mm will be enough for secondary electron formation. Secondary electrons, having small energy, will drift on the direction of electric field to the detector. Selecting an intensity of the electric field, it is possible to achieve an effective transfer of electrons to the drift gap and further to the detector.

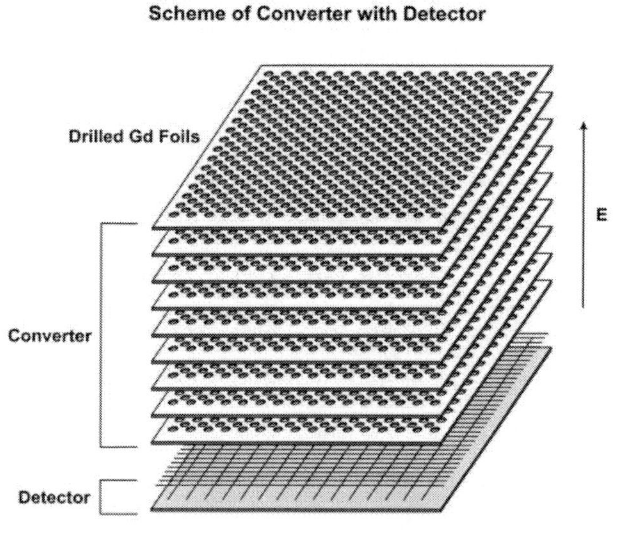

Scheme of Converter with Detector

Figure 38. A schematic sketch of the detector consisting of the complex converter made from drilled by fine step foils located one over other in the same gas volume.

This kind detector will have not so good spatial resolution, which will be modulated with the step of apertures (in this case, 2 mm). For the series of tasks, this resolution is sufficient. In order to improve the spatial resolution, one should prepare foils with finer step, correspondingly, with the smaller diameter of apertures.

Calculations of this kind of collimator are carried out. In calculations, only conversion and Auger electrons with the energy higher than 29 keV isotropic emitting to the all sides were considered. Efficiency for each foil and their sum were separately calculated. In Figures 40 and 42, results of calculations for the converters made from ^{nat}Gd with the thicknesses of 5 and 10 microns correspondingly, are presented. Efficiency is calculated for each foil separately. At absorption of neutrons by lobby foils, a neutron flux to the subsequent foil fall. It results in shielding of foils. The greatest efficiency can be received for foils with the thickness in 1 micron. At use of a foil with the thickness more than 5 microns, the efficiency decreases due to a loss in secondary electron escape. In Figures 41, 43, total efficiency is shown. For calculations is in interest only converters from natural gadolinium as it can increment their total efficiency.

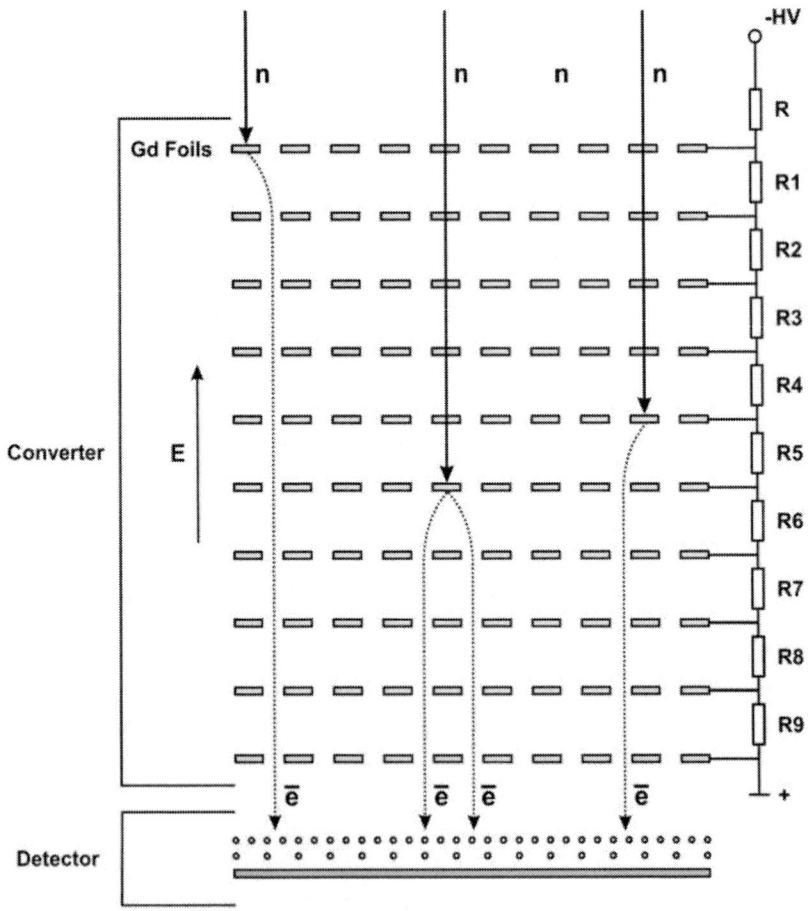

Figure 39. Scheme of the complex detector.

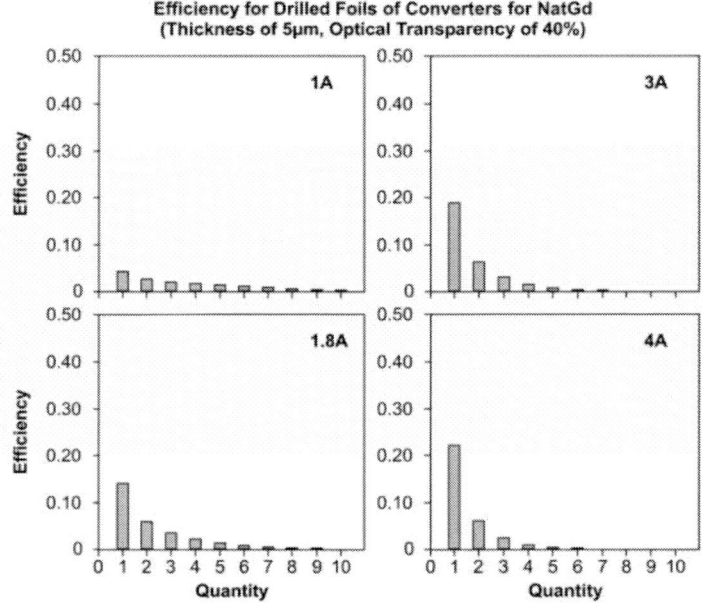

Figure 40. Efficiency of every one of the subsequent drilled converter from 5 microns thickness natural gadolinium with an optical transparency of 40%.

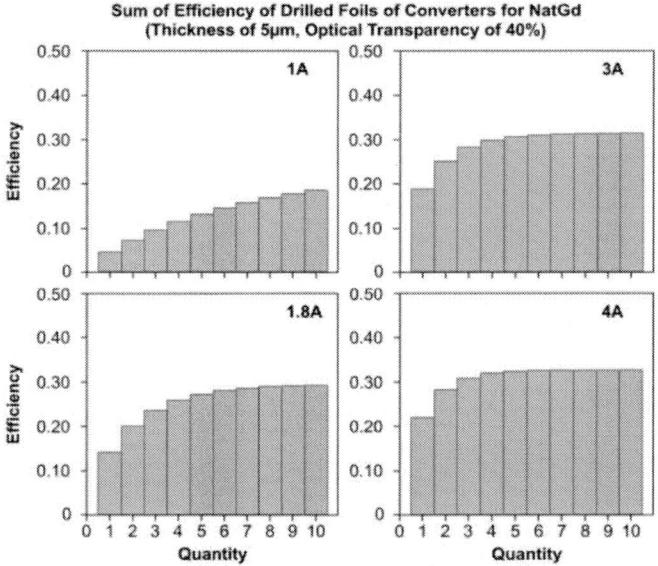

Figure 41. Summary efficiency of the converter consists from 10 drilled foilsr of natural gadolinium with a thickness of 5 microns.

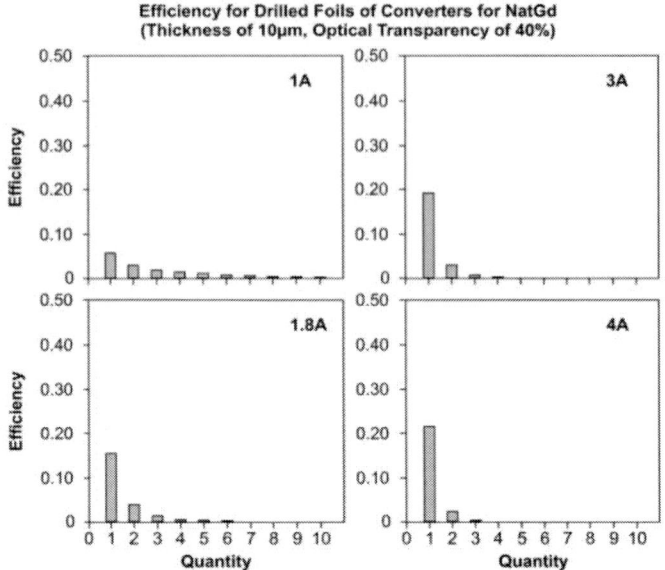

Figure 42. Efficiency of every one of the subsequent drilled converter from 10 microns thickness natural gadolinium with an optical transparency of 40%.

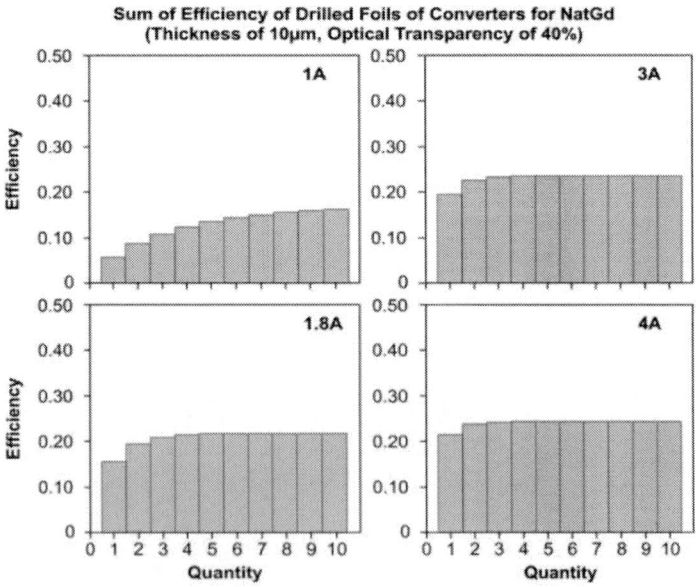

Figure 43. Summary efficiency of the converter consists from 10 drilled foils of natural gadolinium with a thickness of 10 microns.

Chapter 3

POSITION-SENSITIVE DETECTORS OF THERMAL NEUTRONS WITH GADOLINIUM CONVERTERS

The idea of using thin foils to convert neutrons radiation into charged particles and count the conversion products in a solid-state detector was originally proposed by Feigl and Rauch [4,5]. They employed natural Gd and pure ^{157}Gd convertor foils and measured the escape probabilities and the energy distributions of the escaping electrons with Si surface barrier detectors.

Jeavons et al. [6] applied a multiwire proportional chamber (MWPC) for the amplification and imaging of the secondary charged particles and discussed convertor materials other than Gd. To reduce the position broadening due to the large angular spread and finite range of the escaping fast electrons, a "high-density" drift space was used as initial stage of the detector. This stage is essentially a multi-pinhole collimator where along each pinhole an electric field can established to drift the secondary ionization products towards the amplification and imaging stage. However, the finite size of the pinholes limits the position resolution, and the comparably large mass of the collimator increases the γ-ray sensitivity and reduces detection efficiency.

Melchart et al. [8] proposed a multistep avalanche chamber atmospheric pressure for imaging the electrons escaping from a Gd foil. The exponential avalanche growth in the parallel gap preamplification stage makes the detector more sensitive to charges produced near the entrance point of the particle track into the gas volume, thus ensuring a good localization of the neutron interaction in the thin convertor foil. Indeed, they achieved a position resolution better than 1 mm (FWHM). The resolution was therefore found to be pulse-height dependent, and the best spatial resolution was obtained only by restricting the amplitude range and thus the detection efficiency of the detector.

We also have developed and made normal pressure MSAC with converter of natural gadolinium. Thus in operations, we have made detailed studying of performances MSAC. Possibilities are stabilization of operation and influence of separate devices of detectors constructions on performances of detectors studied. As a result, it was possible to refine considerably performances of detectors and to obtain long-term stability in-process detectors.

Breskin et al. [45] makes further step in development of multistep avalanche chambers; there was development of low-pressure multistep avalanche chambers. In one of our publication, possibility of reception of a regime of a preamplification in the low-pressure multiwire proportional chambers also has shown at usage additional plain-parallel gap disposed on top the MWPC [46].

Dangendorf et al. [47] has offered for the first time low pressure MSAC with gadolinium converter. It thus offered to use developing techniques to increase the low-energy secondary electron emission of the convertor foil. Despite the exponential growth of the avalanche in the first multiplication stage, the localization resolution of neutron detectors based on low-pressure multistep avalanche chambers depends on the direction and range of the neutron-induced charged particles. Moreover, for weakly ionizing particles (like the conversion electrons from Gd), the detection efficiency depends on the probability of creating at least one slow ionization electron in the first few hundred micrometer thick gas layer of the preamplification gap. An increase in the number of charges produced close to the convertor surface would improve the localization resolution and the detection efficiency.

Gebauer et al. [44] offered fast and high-resolution hybrid low-pressure micro-strip gas chamber (MSGC) detectors and being developed, which lend themselves for setting up large-area detector arrays.

Abbrescia et al. [48] offered Resistive Plate Chambers (RPCs) widespread, cheap, easy-to-build and large size detectors. Here a technique, consisting in coating the inner surface of the bakelite electrodes with a mixture of linseed oil and Gd_2O_3 reported. This allows making RPCs sensitive also to thermal neutrons, making them suitable to employ for industrial, medical or de-mining applications. Thermal neutron-sensitive RPCs can be operated at atmospheric pressure, are light weighted, have low -ray sensitivity and are easy to handle even when large areas have to be covered.

In Takahashi et al., a new series of experimental imaging plate neutron detectors (IP-NDs) were made, where the composition of the respective IP-NDs, containing a photostimulable BaFBr:Eu^{2+} phosphor and a neutron converter material, Gd_2O_3 or ^6LiF, were varied systematically [49]. The different conversion processes between natural abundant Gd and ^6Li caused distinct imaging properties. The imaging steps and the factors governing the image quality are considered in the same way as X-ray radiography, and the quantum noise is estimated by the effective neutron absorption in these IP-NDs when they read by a BAS2000 image reader.

Gunji et al. [50] have been developing a new type of neutron imaging detector with position resolution better than 10 micrometers by absorbing liquid scintillator for neutron capture into a capillary plate by capillary phenomena. Establishing the methods to absorb liquid scintillator to the capillary plate and attaching it into a photomultiplier, we irradiated the detector with neutrons and alpha particles. From the results of basic experiments, it recognized that the capillary plate absorbing liquid scintillator could operate as a neutron detector.

Bruckner et al. [38] have been developing position-sensitive silicon detectors, and the corresponding electronics allow the construction of fast time response thermal neutron detectors. These detectors also exhibit excellent position resolution by combination of silicon detectors with thin Gd converter foils. Authors constructed several one- and two-dimensional prototype detectors. The position resolution and the detector efficiency for different converters at wavelengths from 1.1 to 3.3 A^0 were determined at the TRIGA reactor in Vienna and at the ILL in Grenoble. Spatial resolutions of less than 100 μm and efficiencies up to 40% have achieved. These detectors can also be used for phase topography experiments using perfect crystal neutron interferometers.

3.1. NORMAL PRESSURE MULTISTEP AVALANCHE CHAMBER

A multistep avalanche chamber (MSAC) consists of a conventional multiwire proportional chamber (MWPC) placed within a single gaseous volume with additional grid electrodes forming preamplification and drift spacing. Gadolinium converter settles down directly ahead of a preamplification grid. MSAC registers electrons escaping off in a back hemisphere. In the preamplification gap, there forms an electron-photon avalanche, which transferred to MWPC for terminal amplification and registration. Due to the exponential amplification in the preamplification gap, there is implemented a tie of coordinates to the entrance point of particles to the detector volume. The spatial resolution of MSAC can be improved by the amplitude analysis and the consequent mathematical processing of events up to 0,2-0,3 μm. The gas amplification coefficient of MSAC is of the order 10^{6-7}, which provides registration of single electrons with practically zero energy. The registration efficiency of the MSAC-based detectors mainly defined of converter's characteristics.

In the Physical Technical Institute of Academy of Sciences of Republic Tajikistan, the detector based on MSAC with natural Gadolinium converter was developed and created. Appearance of the detector is shown in Figure 45.

Figure 45. Appearance of the detector and serving electronics.

The basic parameters of detectors are: sensitive volume - 200x200 mm, gaseous mixture - Ar + (1,5%) of n-geptane, operation pressure - 1 atm, detection efficiency (at $\lambda=1.8A^0$)-35%, spatial resolution-<1mm (FWHM), time resolution (by anode signal)- 20ns, counting rate-$10^6$1/s.

Now operations on studying of detector characteristics and its tests are performed. Studying of performances of detectors frequently makes with usage gamma- and beta- sources. Especially well for these purposes approaches β - a source ^{14}C with the maximum energy of electrons 150 keV. Taking into account energy weakening of electrons in an input window of the detector, the energy distribution of electrons becomes closest to an energy distribution of electrons escaping from the gadolinium converters at response of a radiative capture of neutrons by gadolinium nuclear.

3.1.1. Characteristics of the Multistep Avalanche Chambers

Structurally, MSAC consists of MWPC, located in common gas volume with electrodes A, B and C (Figure 46) on which corresponding potentials brought. These electrodes form conversion (AB), preamplification (BC) and the drift (CD) gaps. In gadolinium converter electrons generated, and only their part can escape a converter material to the conversion gap. In conversion, gap formation of the secondary electrons under the influence of a feeble electric field (~ 2 kV/cm) is carried out. Generated electrons transferred in plain -parallel preamplification gap. In this gap in the strong electric field (up to 10 kV/cm), the electron-photon avalanche develops. Only part of electrons, from the formed avalanche, can reach the drift gap and transfer to multiwire proportional chamber. In MWPC, its further amplification and registration is carried out.

The basic contribution to avalanche development in MSAC is a photo-ionization process. That is, ionization of additive atoms by the quantum which are let out by rare gases at transition in the basic state.

The quantity of the molecular additives in gas mixtures MSAC usually is composed by 1- 4%. At higher concentrations, despite reduction of a free length of secondary electrons and the avalanche, develops less effectively. It is closely associated with reduction of medial velocity of drift of electrons and, accordingly, with reduction of their energy.

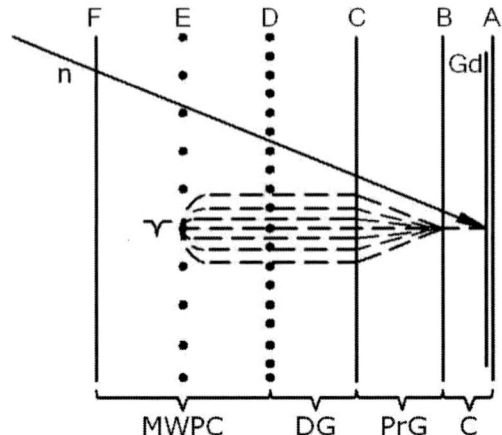

Figure 46. Scheme drawing of the detector, where MWPC-multiwire proportional chamber, DG-drift gap, PrG- preamplification gap, C- conversion gap, Gd- gadolinium foil.

Rare gases and usually argon as basic components in gas mixtures are used. The photo-ionization process of amplification occurs on mixtures of argon with acetone, a bensole, n-geptan, etc. The coefficient of gas amplification MSAC is represented expression:

$$G = k \cdot Gpr \cdot G_{MWPC}, \qquad (47)$$

where: Gpr - and G_{MWPC} - coefficients of a gas amplification of a preamplification gap and MWPC. The coefficient (k) characterizes efficiency of transfer of an avalanche from field with high electric field strength (a preamplification gap) in field with low field strength (the drift gap) and is equal to ~Edr/Epr. The relation of fields in the chamber makes 1/5- 1/10 and only 10-20% of electrons transferred in to drift gap, the quantity of transferred electrons less than the specified value is real and depends on an optical transparency of electrodes. Strength of electric fields in the drift gap and in MWPC, it is approximately identical that allows practically free lost to transfer electrons from the drift gap in MWPC.

Avalanche development process can viewed in the guess of its purely exponential character of development. In preamplification gap in strong (up to 10 kV/cm) electric field, drifting electrons form the electron-ionic steams. Generated electrons in turn participate in the further development of an avalanche.

The gas amplification coefficient viewed by expression [7]

$$Gpr = n / n_0 = \exp(\alpha L), \qquad (48)$$

where, α - coefficient of Tausend, L - a space of a preamplification gap. This expression is valid for a charge formed in the upper stratum of the gap of preamplification. At registration quantum, their absorption occurs along all gap and the amplification gain for each case will spotted as

$$Gpr = \exp(\alpha (L - Xi)), \qquad (49)$$

where: Xi - depth of absorption quantum along a gap.

At registration of the charged particles, primary ionization radiation occurs on all trajectory of a particle, and the gas amplification coefficient in this case spotted as

$$Gpr = (\exp(\alpha L) - 1) / (1 - \alpha L), \qquad (50)$$

The coefficient of preamplification gain of MSAC can reach quantities 10^5 - 10^6, and the full gain coefficient $G = 10^7$. At achievement the critical sizes ($\alpha L = 20$) of avalanche, corresponding to quantity of electrons in an avalanche $\sim 10^8$, the avalanche according to a rule of Reter develops into a streamer. The streamer leads to a spark disruption in MSAC. This rule enters restrictions on greatest possible Gpr. Therefore, at registration of one-electron events, Gpr can make quantity 10^6 -10^7. At registration of events with big specific ionization losses, quantity maximum achievable Gpr decreases, for example Gpr - 10^4 -10^5 at registration gamma -quanta with energy of 6 KeV and $\sim 10^2$ at registration α - particles with energy in some MeV.

3.1.2. Gas Amplification and Efficiency of Registrations MSAC

Measuring of coefficient of preamplification gain (Gpr) for various gas mixtures spent. Gpr was spotted under the relation of quantities of signals from an anode plane at change of a high-voltage voltage on a preamplification gap. Coefficient Gpr can reach quantities 10^5, at Gpr> 10^4 rare spark disruptions in a preamplification gap are observed. With magnification of an amplification values intensity of spark disruptions are incremented. At Gpr> 10^5, the continuous spark discharge localized under a radiation sources is observed.

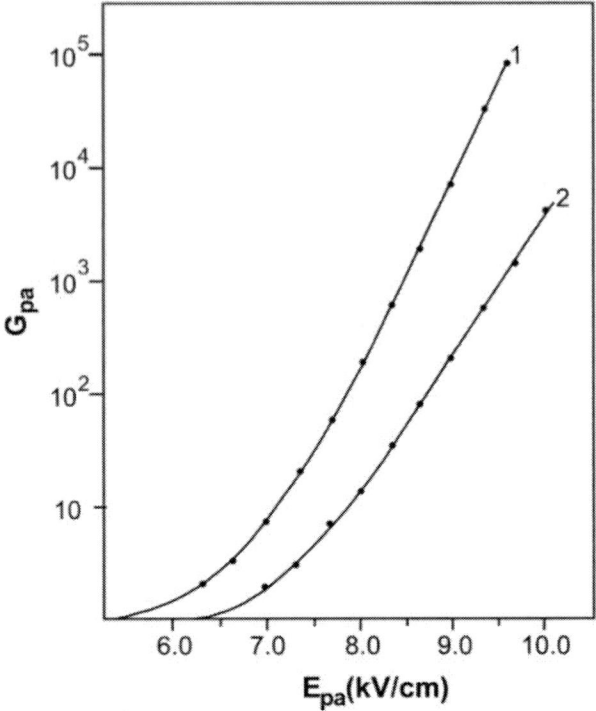

Figure 47. Factor of preliminary amplification in dependence on high voltage on pre-amplification gap. A gas mix of argon with 1,6% of acetone (a curve 1) and 3% of acetone (a curve 2).

Value of coefficient of preamplification gain can reach quantities 10^5, at registration γ-quanta with energy of 6 KeV. It is necessary to note that at quantities Gpr> 10^4, rare spark disruptions on the area in a preamplification gap are observed. At definition of the full coefficient of preamplification gain in a gap, it is necessary to consider losses of electrons while transferring avalanches in the drift gap. For a relation of fields of the drift and preamplification gap 1/5 full value Gpr = $5*10^5$ that corresponds to value of coefficient α = 4,4 mm^{-1}. Common amplification gain of MSAC taking into account amplification in MWPC can reach values G = 10^6 -10^7. At Gpr > 10^6 the recurring operation, MSAC caused by a photon feedback was observed. A part of the ultraviolet quanta that are formed in MWPC can reach a Gd converter, causing occurrence of photoelectrons from materials, impulses from which observed through 1μs after the first occurrence. At Gpr> 10^7, avalanche develops into a streamer that comes to a spark disruption in MSAC.

Examinations efficiency registration and the time resolution spent on a gas intermixture of argon with 2% of acetone. Extent of a plateau counting rates studied at registration γ- quanta ^{55}Fe. In Figure 7 are given extends of plateau counting rates of signals from anode MWPC at various values electric field strength of a preamplification gap depending on quantity electric voltage on the anode MWPC (Ua). The curve 1 on Figure 48 measures at low intensity of Epr, thus an avalanche did not develop and only transfer electrons from conversion to the drift gap, carried out. At Epr = 8- 10 kV/cm quantity Gpr essentially increases that considerably increments extent of a plateau counting rates (curves 2-4). It is visible that the plateau reaches quantities 800 V in at value Gpr 10^4.

Efficiency of registration of charged particles with the minimum ionization losses was explored with β - particles ^{90}St (Emax = 0,53 MeV). In Figure 49(A), observed dates depending on quantity of anode voltage MWPC for the various field gradients in a preamplification gap presented. Measuring spent in the absence of a conversion gap. At enough high values of Epr with efficiency ~ 98%, register the detector β - particles, quantity of a plateau makes 300 V.

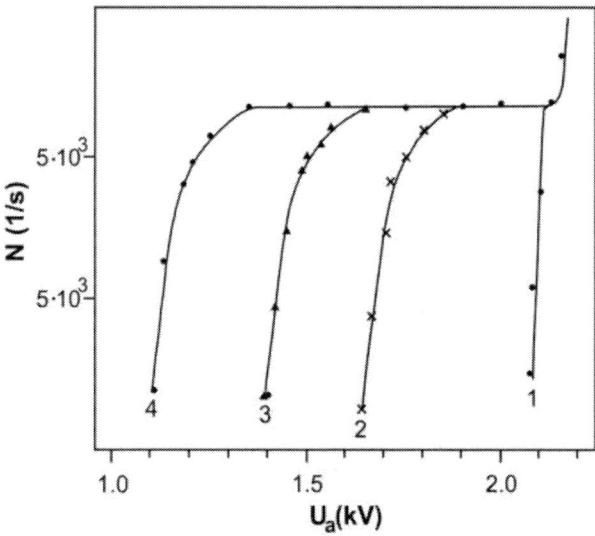

Figure 48. Counting characteristics of the detector at registration γ-quanta ^{55}Fe on dependence on high voltage on MWPC. Value Epr is equal 2 kV/cm (1); 8,3 kV/cm (2); 9,0 kV/cm (3) and 9,7 kV/cm (4).

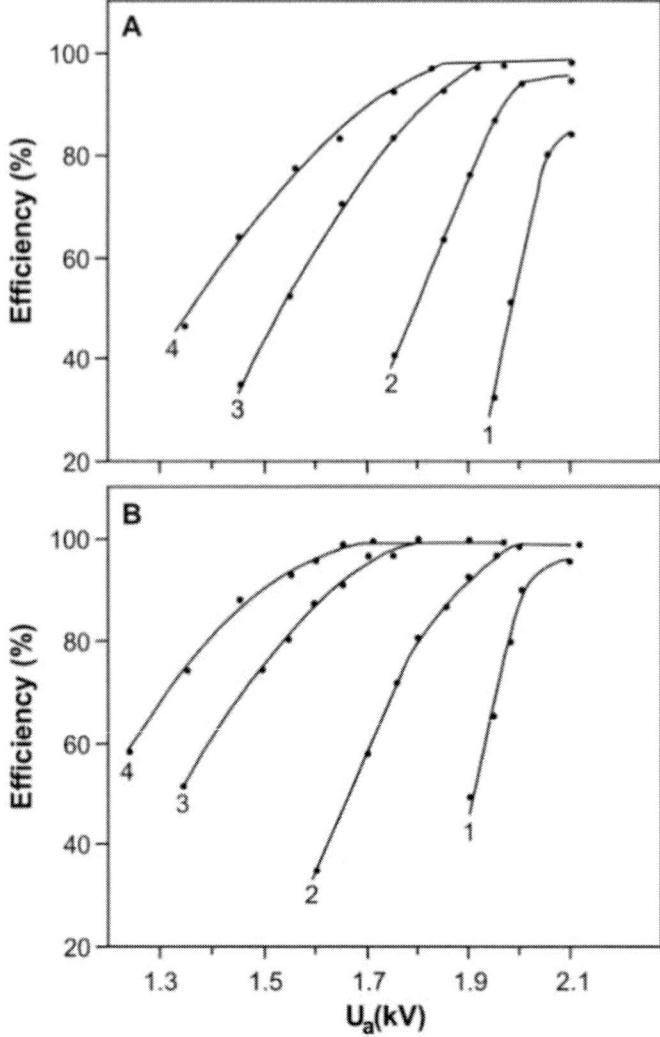

Figure 49. Efficiency of the detector at registration β- radiations depending on voltage on MWPC. Value Epr is equal 8,3 kV/cm (1); 9,0 kV/cm (2); 9,7 kV/cm (3); 10 kV/cm (4). A - the detector without conversion gap; B - size of gap AB = 1 mm.

As it told above, quantity Gpr for the electrons of secondary ionization formed in different points on a thickness of a preamplification gap is various. Introduction of a conversion gap by quantity of 1 mm leads to magnification of a plateau of efficiency of the detector and allows reducing value of necessary anode voltages.

3.1.3. Spatial Resolution of the Multistep Avalanche Chambers

Multistep avalanche chambers (MSACs), as well as plane-parallel avalanche detectors, surpass of the ordinary multiwire proportional chambers (MWPCs), their high accuracy in determining coordinates along both the X – and Y- axes, the presence of the so-called "focusing effect" (i.e., the coordinates of the point at which a particle hits the detector's sensitive volume are measured) and a high gas amplification coefficient.

An electron–photon avalanche that develops in the preamplification gap is transferred through the drift gap to the MWPC for further amplification and detection. Coordinate data read out of the MWPC cathodes via delay lines, and the anode signal can used for amplitude selection of events.

The path of photoelectrons, Auger electrons, and fluorescent quanta limit the spatial resolution of gas-filled position-sensitive detectors (GFPSD) for a narrow collimated γ-ray beam. For example, 5.9-keV γ-rays are converted in argon, mainly on its K subshell with a 3.2-keV ionization potential escaping photoelectrons with an energy of 2.7 keV have a range of 30 μg/cm^2 (i.e., 250 μm of argon at normal pressure and temperature). A fluorescent quantum is produced in 88% of events; its energy is 3.2 keV and its mean free path in argon is ~40 mm. Absorption of a fluorescent quantum inside the detector results in emission of an electron with an energy of 3 keV and a range of 250 μm.

Let us consider the effect that the photoelectron range exerts on the spatial resolution of a GFPSD. Conversion of γ- rays assumed to occur at one point, and detection efficiency for fluorescent quanta thought to be low and can be ignored. Photoelectrons escaping from the conversion point in all directions form a spherical cloud of secondary electrons with radius Re equal to the photoelectron range. The full width at half-maximum (FWHM) of the projection of this cloud onto the plane of the detector is 3/2 R_e. The dependence of the spatial resolution on the electron energy can also found from the empirical formula in [51]:

$$\sigma\ (\mu m) = a\ E_e\ n/\rho, \tag{51}$$

Where E_e [keV] is the electron energy; ρ [mg/cm2] is the specific density of the detecting medium; and the coefficients have the following values:

a= 30, n= 1.30 for E_e= 1–5 keV and

a= 17, n= 1.64 for E_e= 5–20 keV.

Apart from the photoelectron range, the spatial resolution is dependent on other factors, too, particularly on the resolutions of the delay line and the data acquisition electronics.

The spatial resolution of the MSAC was investigated by using a highly collimated (slit width, 50 μm) ^{55}Fe source, which was moved over the detector surface. The experimental results presented in Figure 50. The FWHM was 400 μm. Taking into account the contributions of R_e and the collimator, the intrinsic resolution of the detector was 260 μm (FWHM) [52].

When MWPC detects charged particles from an isotropic emitter, such as Gd converters, its spatial resolution is rather poor. This is explained by the fact that, in the case of oblique incidence, secondary electrons produced along the track of electrons form an extended cloud of charges that collected onto the MWPC anode plane and the center of this cloud are displaced from the particle's hit point through at significant distance. As distinct from the conventional MWPC, gas amplification in the pre-amplification gap of the MSAC obeys the exponential law; therefore, the first electrons produced in the upper layer of the gap were responsible for ~80% of the whole signal amplitude. This improves substantially the spatial resolution of the MSAC for isotropic radiation and gives rise to the so-called "collimating" effect characteristic of the MSAC.

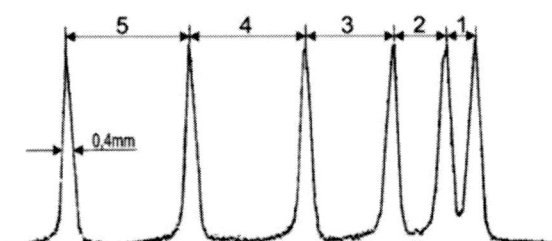

Figure 50. Spatial resolution of the MSAC. The highly collimated 55Fe source moved with a step of 5, 4, 3, 2 and 1 mm.

In the first operations on studying of performances, MSAC with gadolinium converter [8] dependence of the space resolution on amplitude of displayed signals has been detected. The amplitude of a signal is, in turn, out to an angle of an escaping of electrons from a converter material.

The spatial resolution can improve by using amplitude selection of events in electron detection [52]. With this aim in mind, we used two radiation sources (^{14}C) with approximately equal activities, shaped as spots ~1mm in diameter and spaced 1 mm apart (the spacing between the boundaries of these spots). The sources were placed on the entrance window of the detector. Anode signals used for amplitude selection of events by differential amplitude discriminator. The gating signal for data acquisition was generated by the discriminator and arrived at the control input of the time-to-amplitude converter. Information was acquired with a multichannel amplitude analyzer. The amplitude spectrum of the anode signals from the detector irradiated by β-particles of an uncollimated source (^{14}C) are shown in Figure 51.

Particles incident on the surface of the chamber at small angles travel a significant distance in the upper layers of the pre-amplification gap, giving rise to additional avalanches due to secondary electrons along primary particle tracks. These particles are associated with higher-amplitude signals. As a result, as the angle of incidence decreases, the signal amplitude increases, the electron cloud grows in size, and its centroid displaced toward the direction of the primary particle. A small bend on the right of the spectrum can be attributed to the presence of a 0.5-mm-wide conversion gap and represented by the superposition of two spectra, one of which is formed by particles that suffered interaction in the conversion gap and generated secondary electrons and the other of which is due to particles that passed the conversion gap without interaction.

The coordinate resolution obtained when all events were detected (segment AD of the curve in Figure 51) are shown in Figure 52 (curve 1). The ratio of the depth of the dip between the peaks to their height is 0.25. It is possible to improve the spatial resolution by selecting (using the differential discriminator) events corresponding to the portion of the anode signal spectrum from A to B (Figure 51). Incidentally, the ratio of the depth of the dip to the peak height is 0.4 (curve 2 in Figure 52). In this case, the spatial resolution is limited by the own resolution of the detector system, which is dependent both on the delay time per unit length of the delay line and on the accuracy of the electronic channels. When the coordinate information obtained from events with high signal amplitudes (segment CD in Figure 51), the spots cannot be resolved (curve 3 in Figure 52).

The spatial resolution of the MSACs has been investigated. The intrinsic spatial resolution is 260 μm (FWHM) or 100 μm (σ). The spatial resolution of the detector under exposure to soft γ-rays of the ^{55}Fe source is 400 μm (FWHM). The spatial resolution of the MSACs is dependent on the angle of incidence of conversion electrons. The high signal amplitudes due to particles incident on the surface of the chamber at small angles (25% of all events) considerably impair the spatial resolution. By analyzing the signal amplitude and selectively measuring the coordinates, or by estimating the signal amplitude and its contribution to the reconstructed image, it will be possible to raise the spatial resolution when electrons escaping from solid-state neutron (Gd) or γ-ray converters detected.

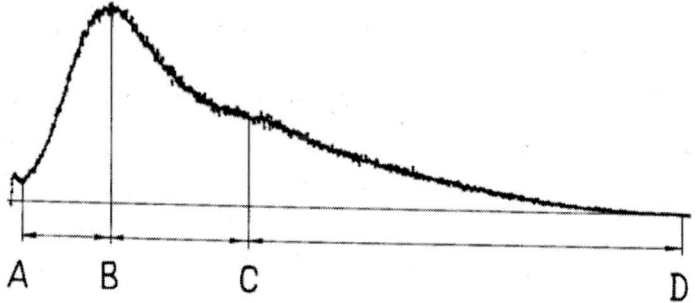

Figure 51. Amplitude spectrum of β particles from the ^{14}C source.

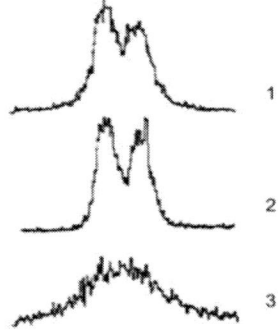

Figure 52. Spatial resolution of two radioactive zones of ^{14}C isotope, each 1 mm in diameter, located with 1-mm spacing between the boundaries.

3.1.4. Influence of Gas Mixes on Characteristic MSAC

Researches of influence on concentration of the additive on characteristics MSAC brought on a gas mix argon + acetone. Acetone as the additive has been chosen that allowed changing in a wide range concentration of the additive in a mix. In Figure 53, values of dependence of factor of preliminary amplification (Gpr) from value of an applied voltage (Upr) for various values of concentration of acetone are brought. From the figure, it is visible that with increase in concentration of acetone at fixed Upr, the factor of preliminary amplification decreases. Thus, Gpr can increase at corresponding increase Upr. Reduction of factor Gpr is connected with reduction of average energy drifting electrons as dominating process becomes interaction electrons with molecules of the additive. This interaction leads to sharp loss of energy electrons. Transition of molecules in the basic state is carried out without radiation, through the rotary and oscillatory transitions. Dependence of amplification on concentration of the additive at the fixed voltage on a preamplifying gap is well described by following expression [53]

$$G(p) = 10^{-(n \cdot pi)}, \qquad (52)$$

where: pi - concentration of the additive in%: n - the factor depending from pi and for a range $0.5 \div 3\%$ is equal 2. So change pi within the limits of $\pm 0.5\%$ cause change G in ± 10 times. The dotted curve designates border sparking in a pre-amplification gap, optimum for achievement of the maximal factor of amplification are concentration $1.5 \div 2\%$.

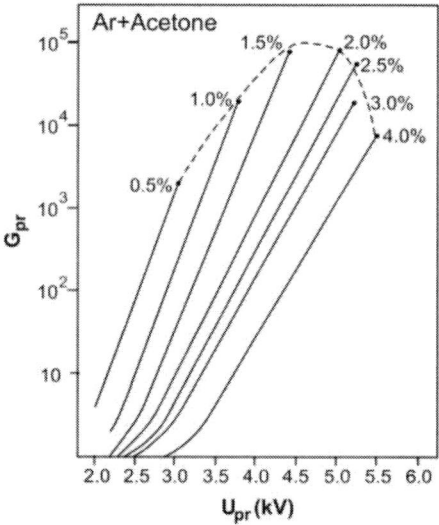

Figure 53. Dependence of factor of preliminary amplification MSAC on a high voltage on a pre-amplification gap for various values of concentration of acetone in argon.

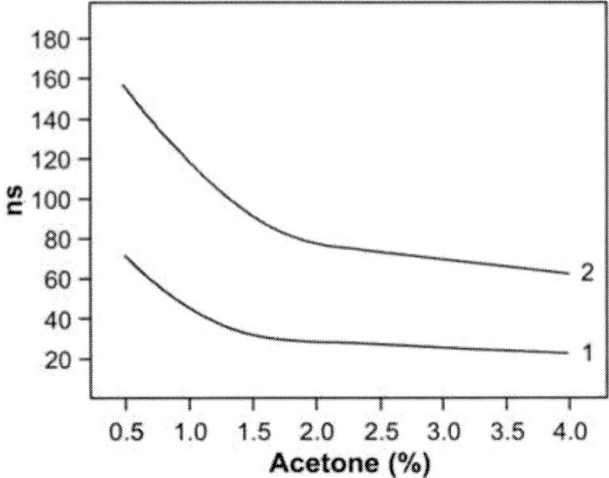

Figure 54. Dependence of time resolution of MSAC on concentration of acetone in a gas mix. A curve 1 - FWHM, a curve 2 - width at a level of 97% of the registered events (FWTM).

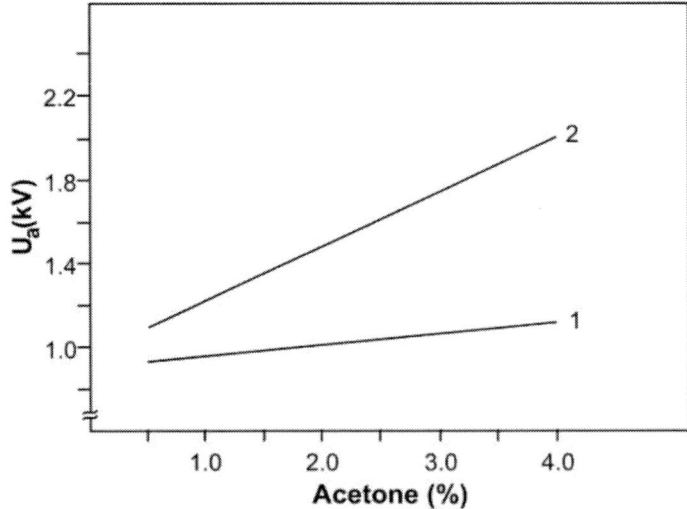

Figure 55. Extent of a plateau counting rate characteristics of MSAC for various concentration of acetone in a mixture. Curves 1 and 2 accordingly, the beginning and the end of a plateau for a source ^{55}Fe.

Influence of additive concentration on the time characteristics MSAC were investigated by β - a source ^{90}St. Measurements were spent at fixed values G and Gpr. In Figure 54, dependences of the time resolution of the detector for a mix argon + acetone are resulted at various concentration of acetone. The curve 1 shows FWMH, a curve of distribution of 2 widths at a level of 97% of the registered events FWTH. From the figure it is visible that with increase in concentration of the additive time resolution of MSAC, which for concentration of 4% makes 21 nanoseconds (FWMH) and 60 nanoseconds at a level of 97% of the registered events, improves.

Researches of influence of acetone concentration on plateau length of counting characteristics was carried out. Researches were spent at fixed factors Gpr. The curve 1 on Figure 55 shows voltage on anode MWPC, corresponding to the beginning of a plateau for a source ^{55}Fe. The curve 2 shows voltage MWPC corresponding to the end of a plateau counting characteristic. From the figure it is visible that with increase of additive concentration, the extent of a plateau, this increase, basically increases, occurs due to displacement of the end of a plateau. Therefore, extent of a plateau makes 900 V for the concentration of acetone equal of 4%.

3.1.5. Ways of stabilization of operating modes for work MSAC

Proceeding from above the characteristic show that the highest factor of amplification can be reached for a mix of argon with 1,5 ÷ 2,5% of the additive. At the same time, the factor of amplification strongly depends on variation of concentration of the additive (in $\pm$ 10 times) at change of concentration of the additive on $\pm$ 0,5%. Other characteristics of MSAC depend strongly on the variation of concentration of the additive as well. In order to receive the long-term stable operating mode MSAC, it is necessary to stabilize concentration of organic compound applied as additives.

One of the possible ways of stabilization of concentration of additives is thermostat control of flasks with a liquid through which gas passed. Thermo stating can be carried out by means of thermostats, or at the temperature of a thawing ice. For maintenance of concentration of known additives within the limits of 1 ÷ 3%, it is necessary to pass part of gas through a flask with a liquid. It dictates necessity of stabilization of gas speeds by means of highly stable reducers. However, even at use of similar reducers, there is a necessity to expose and supervise carefully gas having blown on rotametrs. For the further increase of stability of work of MSAC, it is necessary to use new gas mixes.

The most convenient in operation are organic compounds with low-pressure saturated vapor, which at passage through them of all argon at temperature 0°C allow receiving concentration of the additive in a range 1 ÷ 3%. Such mixes are argon + H-heptanes and argon + isopropyl spirit. At passage of all argon through H-heptanes and isopropyl spirit at 0°C, their concentration, accordingly, makes 1.5 and 1.1%. Received data on these mixes are similar to the characteristics received on mixes argon + acetone.

Other interesting gas mixes for work MSAC are gas mixes on the basis of neon, which emitted more vigorous (in comparison with argon) photons at transition in the basic condition. Energy of photons is sufficient for ionization of some organic additives, in particular, methane.

The gas mixes neon + methane are convenient at operation as allows preparing binary mixes in high-pressure tanks. Researcher of characteristics MSAC on this gas mix were carried out. The mix allows to receive high enough factor of preliminary amplification $G_{pr} = 10^4$. However, at the same time, a neon + the methane mix does not allow receiving a wide plateau. At registration of relativistic particles, extent of a plateau is 100 V for concentration of 2.8% of methane. The plateau of efficiency of registration of relativistic particles can be increased at addition of argon to the neon mixes. Therefore, extent of a plateau of efficiency increases by 100 V for gas mixes Ne + (10 ÷ 15%) Ar+ methane. Neon mixes have higher (two times) time resolution in comparison with argon. So the time resolution for mix Ne ÷ of 2,2% of methane makes 15 nanoseconds (FWHM). Mixes based on neon can be applied to detecting low vigorous electrons in conditions raise gamma – background, first in the neutron detectors with gadolinium convertors.

3.1.6. Dedector Desing

Physically, each MSAC consists of a series of wire electrodes located one after the other in a common gas volume, thereby forming conversion, preamplification, and drift gaps and a MWPC. The schematic diagram of the MSAC is shown in Figure 56.

The conversion gap is formed by convertor *A*, which is made from gadolinium foil and electrode B. The preamplification gap is confined between grid electrodes B and C (these are commercially produced grids or grids composed of orthogonally wind wires). It should be noted that efficient transfer of an electron–photon avalanche from the preamplification gap to the drift gap can be obtained only when the grids have a high optical transparency.

In [54], it is noted that stable operation of the MSAC could be obtained only when the grids were made from thick wires (50–100μm in diameter). For grids composed of 20-μm diameter wires, sparking in the gap makes it impossible to achieve preamplification coefficients $E\text{pr}> 10$. Because drift electrode C begins playing the role of a quasi-anode and development of an electron–photon avalanche becomes a dominant process near the wires of this electrode, which increases significantly, there is probability of a spark breakdown in the preamplification gap.

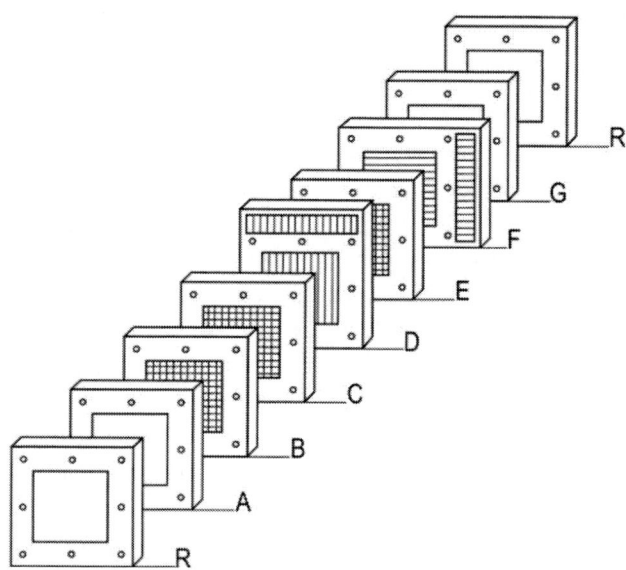

Figure 56. Schematic diagram of the MSAC.

The wire step in the grid is governed by the requirement that the electrostatic field in the preamplification gap be uniform. The electrostatic field assumed uniform at distance $L \geq 3\lambda$, where λ is the grid period. Taking into account that two grids (B and C) form the preamplification gap, it is desirable that the wire pitch selected be $\lambda \leq L/6$, where L is the width of the preamplification gap.

The width of a preamplification gap is usually selected from a range of 3–8 mm. However, the operability of an MSAC with a 20-mm preamplification gap was demonstrated in [55]. The preamplification coefficient $G\text{pr}$ of the

MSAC is exponentially dependent on distance (L) and the first Townsend coefficient (α),

$$G\mathrm{pr} = (e^{\alpha L} - 1)/(1 - \alpha L), \tag{53}$$

and increases with increasing gap width. However, in the case of inclined tracks, the preamplification gap width should reduce to achieve a high spatial resolution. Incidentally, the requirements for the parallelism of the gaps become more stringent.

Stable operation of an MSAC can nevertheless be achieved without using the drift gap. We investigated the characteristics of such a two-step structure. Two-step MSACs also exhibit high gas amplification coefficients, but are not free from some drawbacks. In the case of a spark breakdown, the spark short-circuits the cathode plane, which in turn may lead to a local breakdown of the delay lines and damage the amplifiers.

Creation of the uniform electrostatic field in the preamplification gap is fraught with difficulties due to the cathode discontinuity. MSACs with the inductive readout permit the obtaining of equal spatial accuracies for both coordinates; however, for coordinates along the *Y-axis* parallel to the anode wires, this holds true if an electron avalanche covers adjacent several wires, forming cluster events. An avalanche expands in the preamplification gap in the process of its development and during its transfer through the drift gap in the MWPC.

As a rule, the widths of the drift gap are 5–10 mm. The use of wider gaps requires that a higher voltage be applied to the drift and preamplification gaps and the detector have a larger gas volume.

An MSAC usually comprises a standard MWPC composed of an anode plane and two orthogonal planes X and *Y*. The anode plane is formed by a gold-plated tungsten wire 20μm in diameter, positioning with a pitch of 2 mm. All the wires combined into a common busbar. The cathode planes formed by beryllium- bronze wires, 50–100 μm in diameter, positioning with a pitch of 1 mm and combine into the strips. Each strip comprises four wires and soldered to the delay line. The standard width of the anode-to-cathode gaps is 5 mm.

3.1.7. Effects of Some Constructional Elements on the MSAC Characteristics

The uniformity of gas amplification coefficient G over the detector area is an important characteristic of an MSAC. A difference in the values of G can lead to nonuniformity of the detection efficiency. Figure 57 shows the distribution of G over the MSAC area [54].

Taking the logarithm of Eq. (53), one can obtain deviation of the preamplification gap width
ΔL versus the relative variation in the MSAC amplification coefficient:

$$\Delta L = (1/\alpha)/(\Delta G/G) = (L/\ln G)/(\Delta G/G), \qquad (54)$$

where: G is the mean average of amplification coefficient over the chamber area. This formula helps estimate the deviations from the parallelism in the preamplification gap.

The deviations of the gap width in the MSAC detector from an ideal plane (dashed lines), which are calculated according to Eq. (3), shown in Figure 58. The deviations of mean amplification G are actually dependent both on the parallelism of the preamplification gap and on the spread of the MWPC amplification coefficient.

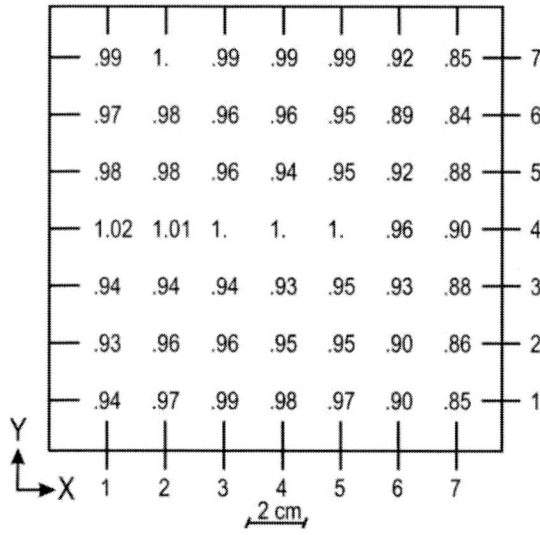

Figure 57. Spread of the amplification coefficient in the MSAC over its area [54].

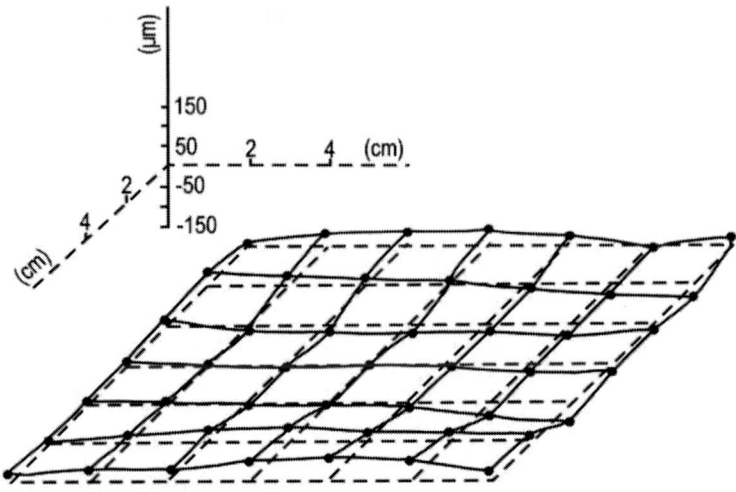

Figure 58. Deviations of the gap width in the MSAC from the ideal plane shown with dashed lines.

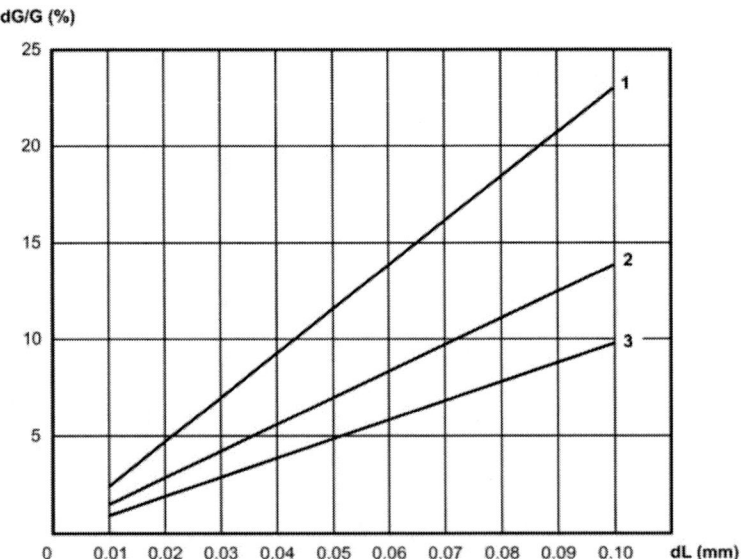

Figure 59. Calculated deviations of the gas amplification coefficient ΔG vs. the spread of the gap with at $G_{pr} = 10^3$ and gap widths of 3, 5 and 7 mm (respectively, curves 1, 2 and 3).

Figure 59 presents calculated deviations of the gas amplification coefficient ΔG such as versus the deviations of gap width from its nominal value at a fixed value of amplification (Gpr= 10^3) and three values of the gap width (3, 5, and 7 mm). From Figure 4, it is apparent that a 100-μm spread of the gap width causes the value of G to vary by 10, 14 and 23% for 7-, 5-, and 3-mm-wide gaps, respectively. An increase in gas amplification coefficient Gpr to a value of 10^5 is followed by an increase in the fluctuations of the gas amplification versus the spread of the gap width. As a result, a 100-μm spread of the gap width causes G to vary by 16, 23, and 39% for 7-, 5-, and 3 -mm-wide gaps, respectively. As the preamplification gap width is reduced, more stringent requirements are specified for the accuracy with which the detector components should be produced. This fact is especially important for manufacture of large-area detectors.

3.2. THERMAL NEUTRON IMAGING DETECTORS COMBINING NOVEL COMPOSITE FOIL CONVERTORS AND GASEOUS ELECTRON MULTIPLIERS

The potential of a thermal neutron imaging system based on a composite neutron convertor foil combined with a low-pressure, multistep avalanche chamber demonstrated in the work [47]. Neutron-induced charged particles from a primary convertor element induce multiple low-energy electrons escaping from a second thin film of high electron-emissive material. Authors investigated the performance of detectors with Gd- and Li-based primary convertors coated with CsI as a secondary electron emitter. It is shown that the detector can be operated with high stability at a sufficiently high gain to detect all escaping particles, fast time resolution, low occupation time and high count rate. A localization resolution of 0.4 mm (FWHM) was obtained. The good imaging performance, free of parallax errors in divergent neutron beams, capability, make this instrument an excellent tool for time-resolved neutron scattering experiments and for neutron radiography and tomography.

The idea of original programmers, therefore, is to replace the single layer convertor with a composite convertor foil; such a foil consists of a traditional neutron convertor coated with a thin film of an efficient secondary electron emitter. Neutron-induced particles from the convertor induce the emission of multiple low-energy secondary electrons, which initiate a well-localized avalanche close to the neutron impact location. To obtain the optimal imaging and timing performance, it is important for the detector to be sensitive primarily to the slow ionization electrons in the vicinity of the convertor foil. These electrons originate either from gas ionization by the fast particle or from ionizations occurring in the foil, leading to the emission of slow secondary electrons into the gas.

Assuming a mean energy of 50 keV for the primary electron escaping from the Gd into the CsI, authors should expect more than 10 secondary electrons emitted on the average from a few hundred µm thick CsI layer. This number is at least five times larger compared to the number of ionization electrons induced in the gas by the 50 keV electron, assuming a 2 mm gap filled with 10 hPa of isobutane. Moreover, due to the statistically distributed ionization processes in the gas and the exponential avalanche growth in the preamplification gap, the effective contribution of these electrons to the final avalanche is even smaller. Thus, the SEE from the CsI is the leading process, and we expect an increase of about an order of magnitude in the pulse-height and detection efficiency close to 100% for fast electrons escaping from the neutron convertor foil.

3.2.1. The Neutron Convertor Foil

The convertor material was deposited on an aluminum disk, 5 mm thick and 75 mm in diameter. A 300 µ*m* thick natural Gd foil was glued onto the Al disk. The Gd surface was coated with CsI layers of various thicknesses, deposited by standard vacuum evaporation. A low-density CsI film of spongy structure was produced by evaporation under 7.5 hPa of Ar. Under this evaporation condition, it is known that the deposited layer has a porous structure and its average density is only a few percent of the bulk density of solid CsI. It is expected that this structure, with its large surface-to-volume ratio, leads to an enhanced secondary electron emission. Unfortunately, authors did not have any control over the effective mass of CsI deposited on Gd, since the thickness monitor did not function under these evaporation conditions.

3.2.2. The Multistep Avalanche Chamber

The preamplification gap is defined by the convertor foil and an 81% transmission grid made of stainless steel. The width of the gap is 2 mm, and it is operated in a proportional amplification mode. After preamplification, the avalanche electrons are transferred to the second multiplication region through a 20-mm-long drift and diffusion gap, which improves the operation stability at high gains. The second amplification stage, a multiwire proportional counter, consists of a 1-mm spaced wire anode plane, 20 μ*m* in diameter, placed between two 1-mm-spaced, 50 μm in diameter wire cathode planes. The localization of the center of the avalanche corresponding to the neutron-capture location in the convertor foil is obtained by analyzing the induced signals on the cathode wires, applying the well-known delay-line readout technique.

3.2.3. Detector Characteristics

To obtain maximum neutron detection efficiency with Gd-based convertors, the avalanche chamber must be capable of detecting events originating from a small number of slow electrons, often-single ones, produced in the vicinity of the convertor surface. To achieve a count rate plateau, the chamber must, therefore, operate at a high gas amplification of several times 10^6. To determine the gain capability of the detector, the neutron convertor foil was replaced by a polished Al photocathode and irradiated with UV light from a Hg lamp. The avalanches were thus initiated by single primary electrons from the photocathode. Close to 100% of the signals clearly separated from the noise. At high gain of about 4×10^6, the detector operates in a stable way over long periods, even in the neutron beam.

For all convertors investigated, the pulse-height distributions and the total detection efficiencies for neutrons at a wavelength of 0.115 nm were measured at various gas pressures and voltages. Although the detector with the pure Gd convertor was operated at significantly higher gain, the pulse-height spectrum obtained with the composite convertor shows a significant increase in the number of events with medium pulse-heights. This reflects the expected increase in the number of slow electrons deposited in the preamplification gap by SEE from the CsI.

The detector with CsI-coated convertors reaches count-rate saturation at significantly lower voltages, which is important for its long-term operation stability. The efficiency values are about 8%, which is in good agreement with the theoretical value for a thick natural Gd foil under backscattering conditions. The detection efficiencies at 40 hPa seem to exceed the theoretical predictions. However, at this pressure, the maximum attainable gain in the preamplification gap is lower by more than a factor of 10 as compared to the value reached at lower pressures. Therefore, the MWPC contribution to the total gain dominates, and the detector becomes susceptible to all charges produced somewhere in the gas, making the detector very sensitive to all kinds of background radiation. For example, γ-rays originating from the reactor or from the Gd foil may be Compton-scattered in the detector wall, the emitted electrons may enter the sensitive gas volume and be multiplied and registered. This is an additional argument in favor of the low-pressure operation of avalanche chambers in this application.

Furthermore, it should noted that for CsI-coated converters, the total conversion efficiency is almost independent of the gas pressure, even at the lowest applied pressures where the probability of producing secondary electrons in the first layers of the gas is negligible. This clearly demonstrates the effectiveness of the CsI secondary electron emission.

A spatial resolution of 0.4 mm (FWHM) was obtained.

In the work, authors present a novel neutron-imaging system based on a composite foil convertor combined with a low-pressure gaseous electron amplification device. The system has many attractive features:

- The concept based on comparatively simple and inexpensive technologies.
- The detector is practically not limited in size or geometry.
- A position resolution of better than 0.5 mm (FWHM) can be achieved.
- In divergent neutron beams, the position determination is free of parallax errors.
- Due to the thin conversion layer (below 100 μm thickness), the arrival time of thermal neutrons can be determined with an uncertainty of less than 100 ns. This excellent time resolution enables time-of-flight and coincidence measurements.
- An operation at very high count rates ($> 10^{-6}$ s^{-1} mm^{-2}) is possible with adequate read-out and acquisition electronics.

The main drawback of the method is the relatively low detection efficiency compared with ^{3}He or scintillation detectors. The efficiency of the detector was found to be in good agreement with the theoretical predictions of the particle conversion and escape probabilities. It is expected that the application of isotopic pure convertor materials will result in detection efficiencies of 36% and 52% for neutrons with wavelengths of 0.1 nm and 0.2 nm with ^{157}Gd convertors. This can be further improved by a multi-foil system, where n foils are read out by $n + 1$ multistep avalanche chambers, recording electrons emitted from both surfaces of each foil.

3.3. HYBRID LOW-PRESSURE (MSGC) NEUTRON DETECTORS

3.3.1. The Detector Principle

A detector scheme with four full-size detector segments of 285 x 285 mm^2 size is shown in Figure 60. Each segment comprises a central composite ^{157}Gd/CsI converter foil, and either side of the converter in 0.25 mm distance extraction grids embedded in 4.5 mm deep low-pressure gas multiplication gaps. The gaps are closed by two-dimensional position-sensitive MSGC plates, which are fabricated by multi-layer deposition on glass plates [44].

The optimal for thermal neutron detection, whereas for cold neutrons, 1–3 μm thick ^{157}Gd converters deliver higher efficiency. The Gd is coated with columnar CsI secondary electron (SE) emitter (SEE) layers delivering a detectable cluster of slow (eV) SEs into the gas volume from a well-defined locus ($\sim$ μm) on the converter surface where the fast CE penetrates. The CsI layers have thickness of smaller than 1 μm in order to keep the sensitivity to energetic γ and X-ray background in the beam low. On the other hand, if photons emitted after neutron capture from ^{157}Gd should emit a cluster of SEs triggering an avalanche above threshold, this signal is as valid and as localized as a CE signal due to the thin converter. Between the converter and the extraction grid, a high field strength of E=5–10 kV/cm is applied for enhancing the SE extraction from the columnar CsI surface.

The extraction gap works in addition as a pre-amplification gap with short mean free path for ionization λ_i in the three-stage gas avalanche multiplication mode used for SE amplification. In this exponential gas amplification mode, only avalanches starting at the converter surface achieve the full gain, whereas avalanches triggered elsewhere in the gas volume remain below the detection threshold. Therefore, the sensitivity to weakly ionizing X-rays and γ- rays will be significantly reduced.

Due to low-pressure operation at p $\sim$ 20 mbar in the very good, quench gas isobutane very high reduced field strengths E/p and thus exponential gas avalanche multiplication can be achieved over three amplification stages:

In the 0.25mm deep pre-amplification gap between converter and extraction grid E/p is as high as 50–500 V/(cm mbar) resulting in λ_i < 53 μm and thus very good time localization of the SE avalanches. Due to the short λ_i, the sensitive detector volume is restricted to the converter surfaces, parallax is avoided and avalanches released from other parts of the detector volume remain below threshold.

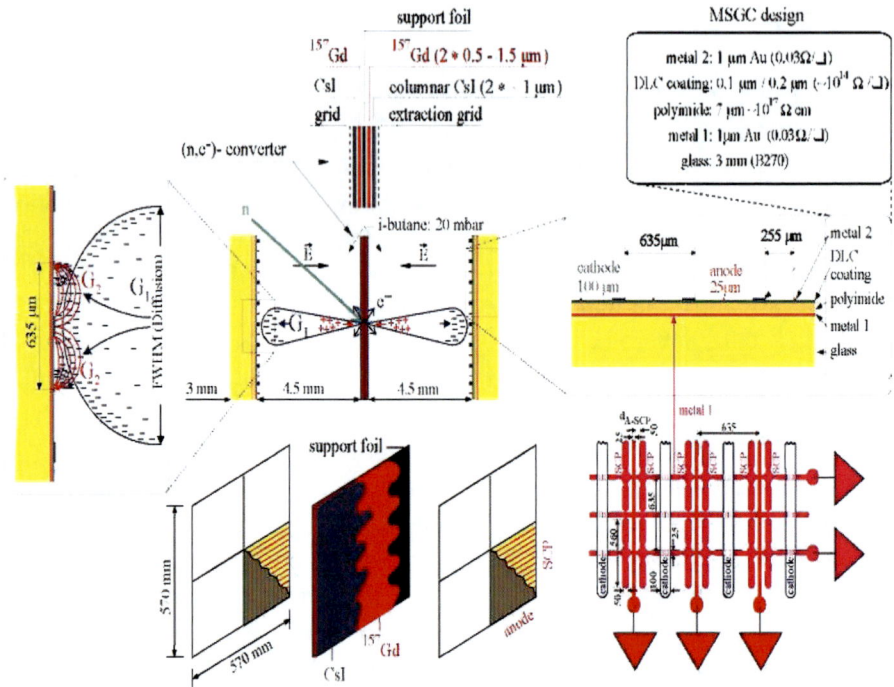

Figure 60. Schematic diagram of a four-segment low-pressure MSGC neutron detector, comprising in each segment (i) a composite ^{157}Gd/ CsI converter foil with extraction grids; (ii) two adjacent low-pressure (p ~ 20 mbar) gas gaps on either side of the converter with high constant reduced electrical field strengths E/p; (iii) two fourfold subdivided MSGC detector planes, which function as amplification and readout elements. The diagram indicates the gas amplification mode at low gas pressures (left inset), the composite converter foil (top), the micro-strip planes either side of the converter (bottom), some coarse details of the micro-strip plate multi-layer design (right) and the layouts (bottom right) of the micro-strip plane (metal 2), with parallel anode and cathode strips, and of the 'Second Coordinate Pad (SCP)' plane with SCP and return current strips (metal 1, magenta).

In the subsequent constant field region extending from the grid to ~ 0.5 mm distance from the MSGC plate, the lower E/p of 100 V/(cm mbar) is still more than sufficient for parallel plate avalanche multiplication. However, only a fraction of 50% of the electrons of the pre-amplification gap can reach this lower-field region by diffusion/scattering through the grid holes. This is enhanced due to the high electron temperature in the high E/p of the preamplification gap and due to the only 20μm thick stainless steel extraction grid with double-sided etched holes of 40 μm diameter and 60 μm pitch. This grid reduces further the UV photon feedback from the end of the avalanche to the converter and thus secondary avalanches.

In the high alternating fields between the micro-strip (MS) anodes and cathodes, micro-strip amplification sets in with rising E/p when the electron avalanches approach the anodes. Due to strong diffusion extending from the pre-amplification gap over the full trajectory length, the avalanche heads are broadened and charge is induced on 3–4 anode–cathode pairs although with Δx= 255 μm a pitch of 635 μm is chosen. Thus, with readout methods determining the center of gravity of the induced charge and with the very good signal-to-noise ratio of the three-stage gas multiplication, a position resolution corresponding to a fraction of the MS pitch is achievable. Due to the higher E/p and shorter drift times, the count rate capability of low-pressure MSGCs is even higher than that one of normal pressure MSGCs, which exceeds 10^6 cps/mm^2, since most of the positive ions are set free in the gas avalanche close to the anode micro-strips and are drifting over a short distance to the cathode strips.

3.3.2. Converter Fabrication

For fabrication of 1-3 μm thick Gd converters at HMI, two half as thick Gd layers were RF sputter-deposited either side on a 6 μm thick Aramid plastic support foil on thin Al barrier layers. Great attention was paid to prepare very uniformly stretched Aramid foils because of the high compressive stresses appearing in the thick Gd layers during deposition. In order to avoid thermal deformation of the Aramid foil, for Gd sputtering, a two-inch sputter source was operated with low power of 50-100 W. In addition, the rear side of the foil was Peltier-cooled via a polished copper plate. For achieving with the small source a homogeneous layer deposition on full-size plates, the substrate must be moved across the source in x and y directions. This is possible in the new chamber but not in the old setup. Therefore, so far converters for prototype size detectors with 101 **x** 101 mm^2 inner frame opening were made. As depicted in Figure 61, columnar CsI layers of ~ 1 μm thickness and with ~ 4-5 times smaller column base width were made by thermal evaporation in an Ar atmosphere of 10^{-3} mbar base pressure, again on thin Al sub-layers.

3.4. RESISTIVE PLATE CHAMBERS WITH GD-COATED ELECTRODES AS THERMAL NEUTRON DETECTORS

Resistive Plate Chambers (RPCs) basically consist of two bakelite plates kept at a 2 mm distance by a grid of plastic spacers [48]. An appropriate gas mixture is circulated in between, and a 4–5 kV/mm electric field is applied. When an ionizing particle crosses the gas gap, subsequent avalanche or streamer processes induce a detectable signal on external readout strips.

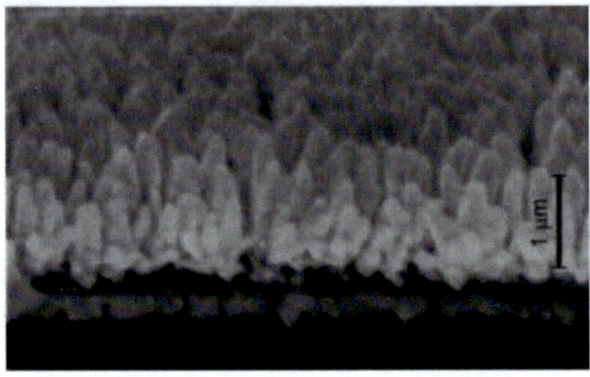

Figure 61. Scanning electron micrograph of a 1-mm thick columnar CsI layer deposited as described in the text.

Figure 62. Appearance of the detector.

Even if this device has been mainly employed to reveal ionizing particles, its possible use to detect neutrons could lead to interesting practical applications like, for instance, the spotting of explosive materials contained in antipersonnel and antitank mines underground. A possible detection technique, accurately studied in the context of the DIAMINE project, uses a ^{252}Cf source placed near the mined ground, which, undergoing fission processes, emits neutrons ranging in energy from about 1 to 4 MeV. Going through the layer of soil overhanging the mine and hitting materials compounding it (mainly H and N), neutrons lose energy in multiple scatterings till termination. The goal is to reveal the backscattered thermal neutrons directing upwards, since an intense enhancement in their number is a precise signature of the presence of the mine. RPCs are very good candidates to fulfill the requirements of an "on-field" application of this technique, because they are cheap and mechanically robust.

Neutron detection can only be achieved after interaction with a suitable material, called converter, which has the role of generating ionizing particles. Two gadolinium isotopes, namely Gd and ^{157}Gd, present in absolute, have the largest cross-section to thermal neutrons (of the order of 10^5 barn).

Cross-section for these converters follows the same typical behavior, decreasing with $1/v$, where v is the velocity of the incoming neutron; in the case of Gd, the cross-section slope is much steeper, starting from an energy around 40 meV to become comparable to others for Ek 1 eV. This means that Gd can be particularly suited to produce detectors specifically designed to reveal thermal neutrons (and not fast neutrons); this is an advantage, since fast neutrons coming directly from the ^{252}Cf source could constitute a sort of undesirable background.

In the work, natural Gd, which is composed of a mixture of many isotopes, of which ^{157}Gd and ^{155}Gd constitute about 30% of the natural composition, has been chosen. This material was actually used in the form of Gd-oxide (Gd_2O_3), which presents itself as a white inert powder, with granules 1-3 mm in diameter. This is inert, very easy-to-handle powder, poses no problem of gas contamination and is very cheap (100$/kg). Gd-oxide powder was put in suspension inside the linseed oil normally coated on the inner surface of the bakelite electrodes, so that the usual method for its polymerisation allows trapping the granules of converter in a solid layer. Since Gd-oxide is not conductive, the electric properties of the electrodes are not altered.

To estimate the possible performance of this kind of detector, suitable calculations have been carried out. In particular, a Monte Carlo program has been developed, taking into account the basic interesting processes, i.e., the interaction of the incident thermal neutron, the emission of the secondary electron with the correct energy spectrum, its travel in a Gd-oxide sheet up to its surface and the eventual entering into the detector active volume.

Total detection efficiency rapidly reaches a maximum of around 10–15μm thickness of about 35%, and after that, it decreases slowly. When the Gd-oxide thickness is lower than 10 μm, the probability of an electron stopping inside the Gd sheet is negligible, and the fast rise in efficiency reflects the increasing number of neutron interactions as the Gd thickness increases. When Gd thickness becomes comparable with electron range, one has to take into consideration the fact that neutron flux decreases exponentially inside Gd-oxide; this means that only a surface layer of Gd mostly contributes to the conversion processes.

Once produced, "forward" electrons have to cross more and more Gd-oxide to exit the sheet, as Gd thickness increases, while "backward" electrons have just a, more or less, fixed thickness to cross. This causes two different behaviors for the curves representing the contributions to the total detection efficiency. In this configuration, one can choose to reveal "forward" or "backward" electrons, depending where the detector active volume is located. Revealing "backward" electrons, one has not to worry much about the layer thickness, which may be difficult to control, provided it is greater than 10 μm.

This new technique was applied to build three RPCs, 10 x 10 cm^2 in dimension, of which one without Gd_2O_3 and the others coated, on the inner surface of one of the two electrodes, with different Gd_2O_3 concentrations.

3.5. THE NEUTRON SENSITIVY IMAGE PLATES

Imaging plates (IPs), two dimensional detectors for ionizing radiations such as X-, β-, γ-ray, ultraviolet light, etc., utilizing photostimulated luminescence (PSL), are used in many fields of applications, such as medical and industrial radiography, autoradiography, X-ray diffraction experiments, and transmission electron microscopy. In each case, the inherent imaging principle is the same explained as follows: after the radiographic image is transferred to an IP, it is scanned point-by-point by a laser beam in an image reader. A series of the PSL emissions corresponding to the scanned pixels is detected by a photomultiplier tube through a high-efficiency light guide to be converted into the electric signals as a function of time. These analog signals are logarithmically amplified and converted into digital signals. By processing these signals through a computer, the computed radiograph can be reconstructed. Moreover, these digital signals can be easily analyzed, stored, and communicated.

Authors have succeeded in experimentally developing an imaging plate neutron detector (IP-ND) [49]. The intrinsic part of the IP-ND is composed of a fine mixture of a PSL phosphor, $BaF(Br,I):Eu^{2+}$, and a neutron converter material, Gd_2O_3 or LiF. The nuclear reactions of Gd with neutrons produce γ-ray of 0.3, 0.4, 1.2 MeV and other energies and internal conversion electrons with an average energy of about 70 keV, and those of ^6Li produce a particles of 2.05 MeV and tritium of 2.74 MeV. These charged particles emitted from the converter can excite the PSL phosphor in the IP-ND, where the ranges of these particles in the phosphor layer are calculated to be less than a few tens of micro-meters, excepting effective penetration length of the γ-ray.

$BaFBr:Eu^{2+}$ is used as a PSL material, and Gd_2O_3 with natural abundance of Gd or ^6LiF is used as a converter material. The average diameters of these powders are 5.0 µm and 3.3 µm or 3.5 µm, respectively. In each IP-ND, 250 µm and 6 µm thick polyethylene telephthalate (PET) films are used as a support and a protection layer, respectively. The size of these IP-NDs is 20 cm X 25 cm or 20 cm X 40 cm. Exposure to neutrons of 2.3 A^0 wavelength was carried out in JRR-2 of Japan Atomic Energy Research Institute (JAERI) and the BAS2000, a bio-imaging analyzer system (manufactured by Fuji Photo Film Co., Ltd.), was used for reading the IP-NDs. The pixel size of the BAS2000 can chosen between 100µ m X 100 µm and 200 µm X 200 µm. The latter was used here.

Image formation process and the image quality factors were achieved by means of the IP-NDs, which is similar to the case of X-ray imaging except for the excitation process through the converter. This consideration is essential for neutron radiography as well, in order to design or utilize the system properly.

FCR7000 is an image reader system used in medical diagnostics (manufactured by Fuji Photo Film Co., Ltd.), where a pixel size of 100 µm X 100 µm was used. The result indicated that the ratio of an X-ray photon noise to a light photon noise was about 8:1 at a spatial frequency of 1 cycle/mm. This means that the light photon noise is a minor component of the quantum noises in this system, which is to say that the number of light quanta detected for a single absorbed X-ray photon is sufficiently larger than unity. As for BAS2000, the relative situation among the quantum noises are not different because the laser energy for a unit area and the efficiency of collecting emitted light are almost equal to those of FCR7000.

In neutron imaging with the IP-NDs, the energies of secondary particles from Gd and ^6Li converters are higher than the average X-ray photon energy of about 40 keV, which corresponds to the exposure tube voltage of 80 kV mentioned above. The number of light quanta detected for a single absorbed-neutron is expected to be sufficiently larger than unity, since the number of light quanta emitted for a single absorbed X-ray photon or neutron can be determined by the energy of the X-ray photon or the secondary particles. Therefore, the light photon noise in the system of BAS2000 and the IP-NDs is estimated to be a minor component of the quantum noises, and the neutron absorption efficiency of the IP-NDs governs the image quality.

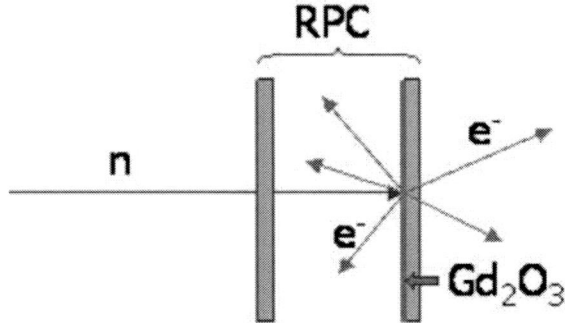

Figure 63. Layout showing the "backward" configuration.

The IP-ND has a severe problem in that it is also sensitive to γ-ray since the PSL phosphor is sensitive to X-rays. To solve this problem, an image processing to subtract γ-ray image from the neutron plus γ-ray image was tried. Since Li emits secondary particles of a much higher energy than Gd, the Li-IPs are not so influenced by γ-ray as the Gd-IPs. It is expected that the Li-IPs are more suited to the imaging where there are higher y-ray to neutron ratios than Gd-IPs.

The PSL is maximized at about 50 mol% of Gd to Ba ratio in the Gd-IPs, while it is not saturated up to 90 mol% of 6Li to Ba ratio in the Li-IPs. In case of the same neutron absorption efficiency, both of the Gd-IPs and the Li-IPs show higher PSL under lower converter concentration. It has been shown that the most important factor governing the image quality is effective neutron absorption in both of the Gd-IPs and the Li-IPs. For the most part, the Gd-IPs can absorb neutrons more efficiently than Li-IPs. However, the amplification rate in the conversion and excitation process of the Gd-IPs is about 30 times lower. Therefore, the Li-IPs are expected to be more suited to the imaging where there are higher γ-ray to neutron ratios than Gd-IPs.

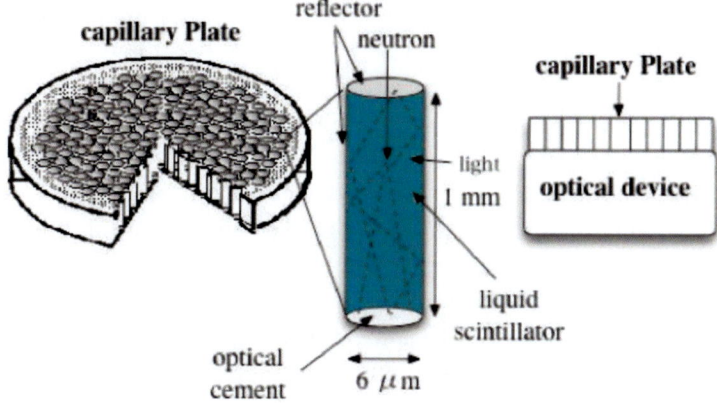

Figure 64. The left figure is the schematic view of the capillary plate. The capillary plate consists of a bundle of fine glass capillaries. The middle figure is the schematic view of one glass capillary. The length and the diameter are about 800 μm and from 6 μm to 100 μm, respectively. Liquid scintillator for neutron capture is absorbed into each capillary. The lights emitted via neutron capture are reflected on the inner surface of each capillary and then are led to a optical device. The right figure is the schematic view of the neutron detector coupled with optical position-sensitive device.

3.5. Neutron Imaging Detector Using Capillary Phenomena and Liquid Scintillator

Neutron beam and the imaging detectors are useful to investigate the inner feature of human body. So several imaging detectors for neutrons such as an Imaging Plate or scintillating-fiber were developed. In general, three important characteristics are required for neutron imaging detectors. These are fine position resolution, timing resolution, and low background. Though the Neutron Imaging Plate has good position resolution of 25 μm and has high detection efficiency for neutrons, it cannot obtain timing information. In particular, the capillaries filled with liquid scintillator will be good candidate for neutron detector with both fine position resolution and good timing resolution [49]. It is because bundle of capillaries with the diameter of 6 μm can be manufactured at present and the liquid scintillator with Gadolinium (Gd) or Boron (Gd) has high detection efficiency for neutrons. So authors have been developing the neutron detector, which consists of liquid scintillator with Gd or B and a capillary plate commercially manufactured by Hamamatsu Photonics Inc. Figure 64 shows the schematic view of the capillary plate and the neutron detector under development. The capillary plate consists of a bundle of fine glass capillaries with uniform length. The diameter and the length of each capillary are 6-100 μm and 800 μm, respectively. Though the effective area of the typical plate is 20 mm in diameter, it is possible to manufacture the capillary plate with larger diameter. Because the plate is thin, it does not have high-detection efficiency for gamma rays, which are background for neutron detectors. On the other hand, it has good detection efficiency for thermal neutrons in spite of the thinness because Gd or B has high-detection efficiency for thermal neutrons. Under the capillary plate absorbing liquid scintillator, optical detector with fine position resolution such as IICCD (Image Intensified CCD camera) or EBCCD (Electron Bombered CCD camera) is attached. As the neutrons are captured by Gd or B, electrons or alpha particles are emitted. They run in the liquid scintillator and the scintillation lights are emitted. If the lights emitted from one capillary by neutron capture can be confined in the capillary and are led to the mounted optical device, the detector can obtain the position resolution of 6 μm.

The key technologies to manufacture such neutron detector are as follows. 1) It is required to absorb the liquid scintillator into each capillary uniformly and to seal it. 2) It is necessary to obtain enough light output by the neutron capture to detect with the optical device. 3) It is required to confine the emitted lights by neutron capture in the capillary and to lead them to the optical device. To realize it, it is necessary to coat the inner surface of the capillaries with light reflector. So we were challenged to establish the first technology using capillary phenomena. Then by irradiation with neutrons, we investigated the light output from the capillary plate attaching a photomultiplier. Finally, we carried out irradiation experiment of alpha particles attaching the capillary plate with a position-sensitive photomultiplier. In this paper, we will report in detail the methods to absorb the liquid scintillator into the capillary plate and results for the basic experiments of neutron and alpha particle irradiation.

Five kinds of liquid were used for these tests. These are water, ethanol, benzene, xylene, and liquid scintillator (BC525) of 2% Gd doped. Authors used two kinds of capillary plates with the diameters of 100 μm and 6 μm, which the side surface inside each capillary is not coated. Though liquid cannot be uniformly absorbed into capillary plates by the method of Type (A) for any combination, we succeeded in absorbing any liquid into capillary plates by the method of Type (B).

Authors have been developing the neutron detector with fine position resolution by absorbing liquid scintillator with Gd or B and to the capillary plate by capillary phenomena. They have established the methods to uniformly absorb the liquid scintillator to each capillary and to seal both sides. Injecting neutrons to the capillary plate attached on the PMT, the basic characteristics have been investigated. From the results, it was confirmed that the capillary plate with the liquid scintillator of BC525 can operate as the neutron detector, though the light output is not yet enough. Moreover, injecting alpha particles to the capillary plate attached on the PSPMT, prototype position-sensitive detector has been tested. However, the position resolution is not good yet because the lights emitted from the capillary are not confined in the capillary. In near future, we will try to manufacture the capillary plate with the liquid scintillator of Boron doped and will establish the method to plate the inner surface of capillaries with light metal for the light reflector.

3.6. POSITION-SENSITIVE DETECTION OF THERMAL NEUTRONS WITH SOLID STATE DETECTORS (GD SI PLANAR DETECTORS)

The central concept of these new detectors is to combine position-sensitive Si planar sensors with external Gd converter foils [38]. The thermal neutrons are absorbed in the Gd foil and the resulting conversion electrons detected in the silicon devices. Figure 65 shows a schematic drawing of a Gd-Si neutron detector consisting of one Gd converter foil and two Si detectors. A position-sensitive Si planar detector consists of an array of many p^+n-diodes on one single substrate. Each p^+ n-junction works as a detector for ionizing particles. The position of the detected particle is given by the individual address of the diode that collects the charge. Each diode is equipped with an individual amplifier chain to ensure real time readout. Typical wafers have a diameter of 4-6 in. and a thickness of about 300 μm. The single structures can be as small as 10 μm and are limited in their extent only by the size of the wafer. A position resolution of less than 10 μm has been achieved with these devices, as well as the construction of large detectors with areas up to m^2, for several high-energy physics experiments.

Considering only one converter layer (no sandwich detectors), the (n, γ) reaction in Gd, which produces also a reasonable number of conversion electrons, is the only suitable one for high detector efficiency. The proton, α-particle and triton from the ^{6}Li and ^{10}B reactions have a very short range. Therefore, these converters can only be used in situations that require low detection efficiency, e.g., beam monitors, or for the detection of cold or ultra cold neutrons. One can also think of doping the Si detectors with Li, but the usual doping concentrations of about $10^{16}/cm^3$ are much too low for efficient absorption of thermal neutrons.

The (n,γ) reaction in Gd produces a quite complex conversion electron spectra with significant energy lines between 29 and about 200 keV. Due to the high absorption, cross-section of Gd the converter can be made very thin, so that the range of the conversion electrons is sufficient to reach the Si-detector. The band-gap in Si of 1.11 eV results in an average energy of about 3.6 keV for the electron hole creation. Therefore, the device must be capable to measure a charge of less than 5000 e⁻ to detect the low-energy conversion electrons, which have already lost part of their energy when escaping the converter foil.

The detectors are made of high resistivity silicon in planar technology and operated in fully depleted mode. For a typical detector, less than 100 V applied to the p⁺ n-junction ensures that the whole detector works as one large depletion region. This mode decreases the detector noise, because it reduces the capacity between strip and backplane, and the thermally generated leakage currents. In addition, it guarantees that the entrance window on the rear side is as thin as the one on the front side and that there are no dead sections between the strips. While CCDs are made of only partly sensitive shift registers and contain inefficient channel stops to separate the columns, Si planar detectors are continuously efficient devices. Another important distinction between these detectors and CCDs concerns their readout. CCDs are integrating devices with charge transfer that means serial readout. Therefore, high-resolution time information is very difficult to obtain. Contrary Si detectors are constructed to measure single events with parallel readout electronics. Provided the lower level discriminator is set sufficiently high, the energy distribution of the conversion electrons does not affect the detector response, as in the case when the spectra would be detected by an integrating device such as film or CCD.

The theoretical time resolution of the Gd Si detector is given by the charge collection time in Si, the transit time of the thermal neutrons through the converter foil and the lifetime of the excited Gd atoms after the neutron capture. None of these intervals is longer than some nanoseconds, which allows the construction of a very fast detector.

For parallel readout, it is necessary to equip each detector diode with the full amplification and readout electronics (more sophisticated readout techniques can reduce the number of channels). For many channels and/or a high position resolution, this can only be done by the use of very large-scale integrated electronics. Although the technique used for the detector fabrication from a general point of view similar to the one used for electronic circuits, it differs in some essential aspects, which prevent so far the construction of detector and electronics on the same wafer.

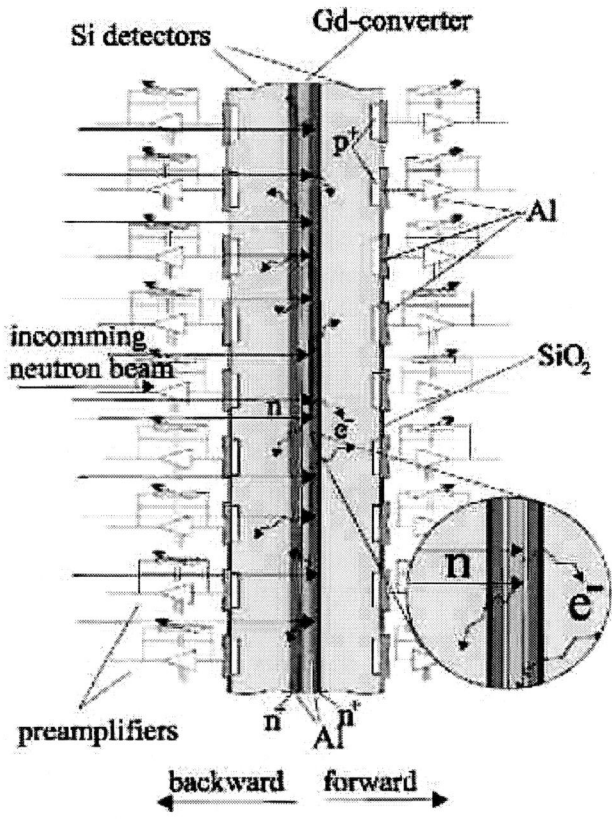

Figure 65. Schematic of a 4% Gd-Si-neutron detector consisting of one converter foil and two Si-detectors with corresponding electronics.

3.6.1. Characteristics of Detector

For every stopped neutron, the probability that an electron will be emitted is given by the conversion factors of the corresponding (n, γ) reaction. As the decay pattern of Gd is quite complex, some simplifications have to be made for the calculations of the detector efficiency. Taking the values of the dominant isotopes ^{155}Gd and ^{157}Gd from the nuclear data sheets, we used only the lowest energy levels which have the highest conversion factors, E_y (keV) =

88.97, 199.21, 296.53, 263.58 for 156Gd and E_y (keV) = 79.51, 181.93, 277.54, 255.66 for 158Gd. The energy distribution of the conversion electrons was calculated by subtracting the binding energies of the K, L and M shells. Although each conversion electron is accompanied by several γ-rays as well as X-rays and Auger electrons, which follow the ionization of the atom, only the conversion electrons and the Auger electrons with an energy of more than 20 keV have been taken into account. For all electrons, which enter the Si detector with a remaining energy of at least 15 keV, the efficiency of the Si detector can be assumed to be 100%.

The efficiency, defined as the number of registered neutrons divided by the number of incident neutrons, was measured in backward direction for six different wavelengths in comparison to a calibrated ^{3}He counter. The results for natural Gd and an enriched (90.5%) ^{157}Gd converter are shown in Figure 11. For thermal neutrons, authors can conclude an efficiency of more than 35% for the enriched ^{157}Gd converter and about half of this value for natural Gd. One can clearly see the increase towards higher wavelengths.

The measured values are constantly higher than the calculated ones; this is caused by X-rays and γ-rays following the neutron capture, which are partly detected. The X-ray lines is more dominant for natural Gd. Conversion electrons, γ-rays and X-rays are emitted simultaneously compared to the time resolution of the amplifier of about 1 μs, therefore, one captured neutron can cause only one signal in the detector. Nevertheless, the electromagnetic radiation can increase the overall efficiency in the case when the electron not detected. Otherwise, the detection of γ-rays and X-rays causes only a spread of the energy spectra.

With the first measurements about three years ago, authors could demonstrate that a one-dimensional neutron detector can be constructed from a single-sided strip detector. The position resolution achieved with this early system is compared with the results of a pad detector and the latest measurements with a double-sided strip detector.

As for the previous setup, the position resolution was determined by measuring the detector response to an edge. An Lorentzian edge spread function was fitted to the data near the edge, which gives a position resolution of FWHM= 210 μm.

The measurements confirmed that the distance between Gd converter foil and detector is the critical parameter for the position resolution. To take advantage of the extremely high resolution of the strip detector, the Gd converter has to be placed onto the detector surface or deposited directly on the Si wafer.

CONCLUSIONS

Theoretical bases of conversion electrons formation are considered at reaction of radiating capture of neutrons;, the value of α depends on four parameters: (l) on the charge of radiating nucleus, (2) on the energy of nuclear transitions, (3) on the nuclear shell from which an orbital electron was emitted and, at last, (4) on multipolity and parity of nuclear transitions.

Based on Bloch theory brake ability of electrons is considered. For electrons, full brake ability usually is divided on two components: a - the brake ability caused by collisions ("collision stopping power"), - mean energy losses by the unit of length of path due to non-elastic Coulomb collisions with the bond electrons of the substance, resulting in ionization and excitation; - radiating brake ability ("radiative stopping power") - average losses of energy on the unit of length of path due to emission of brake radiation in the electric field of nuclear nucleus and nuclear electrons. Division of brake ability into two components is expedient for two reasons. First, methods of their determining are completely different. Second, the energy going on ionization and excitation of atoms is absorbed in the substance rather close to a track of a particle, while the basic part of energy lost in the form of brake radiation leaves far from a track before being swallowed up. This distinction is important, when the attention to the energy "transferred locally" to substance along a track is accented, in contrast to the energy lost by an incident particle. Actually, the share of energy lost in ionization impacts turns to kinetic energy of secondary electrons and is transferred thus to some distance from a track of an initial particle.

Model calculations of efficiency of registration of thermal neutrons by the foil converters made from natural gadolinium and its 157 isotope were carried out. Processes of neutron absorption in the material of a converter and the probability of secondary electron escapes were examined. Calculations were made for converters with the various thicknesses. We have chosen the most optimal converter thicknesses, both from natural gadolinium and from its 157 isotope. While using converters from natural gadolinium, it is possible to obtain total efficiency of 10, 21, 26, 30%, correspondingly, for the neutrons with wavelengths of 1, 1.8, 3 and 4 A^0 with converter thickness of 24, 7, 5, 4 microns. For the converters with 157 isotopes, it is possible to reach total efficiency of registration up to 27, 45, 49, 52% for the neutrons with the wavelengths 1, 1.8, 3 and 4 A^0 with the thickness of converter 5, 3, 2, 2 microns.

In earlier calculations, we took into account only electrons with energy higher than 29 keV. Thus, the low-energy Auger electrons were not taken into account; these are the Auger electrons from L-shell with the energy 4.84 keV and Auger electrons of M-sell with the energy 0.97 keV. These electrons have rather small free path length in gadolinium; these are 0.3 microns (4.84 keV) and 0.04 microns (0.97 keV). They bring the small contribution to the general efficiency at use of converters made from natural gadolinium as the length of free path of neutrons in natural gadolinium makes tens micron. At the same time, their contribution becomes essential at use of converters made from 157 isotope of gadolinium as the free path length of neutrons in them does not exceed 2-3 microns, and this length becomes comparable with length of electron paths.

Calculations of the efficiency of registration of thermal neutrons by the foils made from natural gadolinium and its 157 isotope were carried out. In calculations, the low-energy electrons were taken into account. The received results are in a good agreement with the experimental data.

We have estimated the contribution of X-ray and low-energy gamma-quanta absorbed directly in a material of the converter and resulting in occurrence of secondary electrons. At practical calculations, it is also necessary to take into account the absorption of quanta in materials of detectors. At such calculations, total efficiency can appear higher, as the part of electrons can formed from quanta in a material of the detector. In our calculations, we take into account only an output of electrons from a material of the converter. In case of the account of the contribution of electrons formed by X-ray quanta, the efficiency increased a little, but their contribution is insignificant. The occurred increase is no more than by 1%. It is rather possible that X-ray quanta can affect essentially to the total efficiency at the account of their absorption in a material of detectors. Therefore, gas detectors with Argon filling can increase the efficiency essentially. If the conversion of

Conclusions 113

quanta will occur in a working body of the detector, the full gathering of formed secondary electrons will take place. They can render even greater influences in the case of use of semi-conductor detectors.

Calculations of complex converters representing a set of thin gadolinium foils located on the both sides of supporting kapton foils carried out. These calculations are of interest from the point of view of calculation of GEM structures. They have the similar geometry, but only contain apertures in 300 microns diameter and with 300 microns step. Thickness of the kapton foil considerably influences on the efficiency of neutron registration. With the increase of thickness of a film the probability of the secondary electron escape decreases. These kinds of converters could give a good result at application of natural gadolinium.

The new solid-state converter of thermal neutrons was offered. The converter will consist of a set of thin gadolinium foils located one over other in a gas volume. Foils there will be drilled with the fine step (2 mm) with diameter of apertures 1 mm. These foils will have an optical transparency of 40%; correspondingly, gadolinium will fill 60% of a surface. Secondary electrons will emit for all sides, and in an electric field, they will be soaked up in apertures and further drift in the direction of the detector. As a detector, there are various gas detectors, such as the multi-wire proportional chamber, multi-step avalanche and multi-strip detectors, and so on, that can be used. Calculations made for this complex converter representing a set of thin gadolinium foils located one over other in a gas volume were made. Similar converters can give good result at application of foils from natural gadolinium. Technologically, foil thickness should be more than 5 microns. Thinner films are difficult to drill.

In the article, position-sensitive detector of thermal neutron radiation was described. This detectors use the solid-state converters such as normal-pressure multistep avalanche chambers (MSAC), low-pressure multistep avalanche chambers (LPMSAC), microchannel plates, thin layer scintillators. New types of detectors are offered and developed, such as hybrid low-pressure micro-strip gas chamber (MSGC), position-sensitive silicon detectors (Gd Si PD), resistive plate chambers (RPCs), imaging plate neutron detectors (IP-NDs), liquid scintillator into a capillary plate, and so on.

Basic characteristics of MSAC were described. The factor of preliminary amplification of MSAC can reach value 10^5. However, for reception of a stable mode without sparking, it is necessary to choose it $<10^4$. The full factor of gas amplification of detector G makes $10^6 \div 10^7$. At $G > 10^6$ registration of individual photoelectrons is possible. Extent of a plateau counting characteristic at registration gamma-quantum's with energy 6 keV makes 800 V. MSAC extent of a plateau by efficiency 300 V possesses efficiency of registration of relativistic particles close to 100%. The energy resolution at registration gamma-quantum's from a source ^{55}Fe is equal 18%. The time resolution at use of mixes based on argon 20 ÷ 30 nanoseconds (FWHM) and 15 nanoseconds (FWHM) at use of neon mixes. It is measured, its own spatial resolution of the detector equal of 260 microns (FWHM) and 100 microns (σ). Spatial resolution MSAC at registration gamma-quantum's with energy 6 keV about 400 microns (FWHM). The opportunity of improvement of spatial resolution MSAC in case of registration non-collimated charged radiation by peak selection of events is described. Thus on amplitude of signals, it is possible to judge in the indirect images angle of an input of particles in volume of the chamber. Application of the given way - "electronic collimation" - allows improving twice approximately the spatial resolution. The way can be applied at registration conversion electrons, let out by converters at registration neutron and gamma-radiations.

Dependence of characteristics of MSAC on concentration of molecular additives were investigated. So change of concentration of additives on ± 0,5% causes change G in ± 10 times. Extent of a counting rate plateau strongly depends on change of concentration of the additive as counting characteristics and time resolution of MSAC, too. Ways of stabilization of operating mode MSAC, which at use of the above-stated mixes allow reaching long-term stability of work of the detector, are developed.

Mathematical modeling of this work has been accomplished under the Project T-1157 of the International Science and Technology Center. I would like to thank Professor H. Rauch and Doctor V. Dangendorf for interest in our work.

REFERENCES

[1] Convert, P; JB. Forsyth, (eds.), *Position-sensitive detection of thermal neutrons*, Academic Press, London, 1983.

[2] Meardon, BH; Salter, DC. *A Survey of Position Sensitive Detectors and Multi-Counter Arrays with Particular Reference to Thermal Neutron Scattering*, 1972, RHEL-R-262, 91.

[3] Crane, TW; Baker, MP. Chapter 13, "Neutron Detectors," in Passive Nondestructive Assay of Nuclear Materials, edited by TD; Reilly, N; Ensslin, HA. Smith, *US Nuclear Regulatory Commission* NUREG/CR-5550, March 1991.

[4] Rauch, H; Grass, F; Feigl, B. Ein neuartiger fur langsame neutronen, *Nuclear Instruments and Methods in Physics Research*, 1967, 46, 150-153.

[5] Feigl, B; Rauch, H. Der Gd-neutronenzahler, *Nuclear Instruments and Methods in Physics Research*, 1968, 61, 349-356.

[6] Jeavons, AP; Ford, NL; Lindberg, B; Sachot, R. A new position-sensitive detector for thermal and epithermal neutrons, *IEEE Transaction on Nuclear Science*, 1978, Vol.NS-25, No 1, 553-556.

[7] Charpak, G; Sauli, F. The multistep avalanche chamber: a new high-rate, high-accuracy gaseous detector, *Phys. Letters*, 1978, v. 78B, № 4, 523.

[8] Melchart, G; Charpak, G; Sauli, F. The multistep avalanche chamber as a detector for thermal neutrons, *Nuclear Instruments and Methods in Physics Research*, 1981, 186, 613.

[9] Abdushukurov, DA; Djuraev, AA; Evteeva, SS; et al. "Model Calculation of Gadolinium- Based Converters of Thermal Neutrons." *Nuclear Instruments and Methods in Physics Research*, 1994, v. B 84, 400.

[10] Abdushukurov, DA; Abduvokhidov, MA; Bondarenko, DV; et al. "Modeling the registration efficiency of thermal neutrons by gadolinium foils." *J. of Instrumentation, 2007*, JINST, 2, P04001, 1-13, Archives of Los Alamos National Laboratory USA, 2007, 1-19. http://arxiv.org/ftp/physics/papers/ 0611/0611225.pdf.

[11] Abdushukurov, DA; Bondarenko, DV; Muminov, Kh.Kh; Chistyakov, DYu. "Contribution of nano-scale effects to the total efficiency of converters of thermal neutrons on the basis of gadolinium foils." *Archives of Los Alamos National Laboratory USA*, 2008, 1-9. http://xxx.lanl.gov/ ftp/arxiv/papers/0802/0802.0401.pdf.

[12] Gusev, NG; Dmitriev, PP. Quantum radiation of radioactive nuclear, Moscow, *Energoatomizdat*, 1977.

[13] Kozlov, VF. Manual for Radiation Safety, Moscow, *Energoatomizdat*, 1987.

[14] WWW Table of Radioactive Isotopes, Radiochemistry society, http://www.radiochemistry.org/periodictable/frames/isotopes_lbnl/.

[15] Under edition I.K. Kikoin, The table of physical parameters, Moscow, *Energoatomizdat*, 1976.

[16] Evaluated Nuclear Data File (ENDF), *IAEA-NDS*, http://www.nndc.bnl.gov/ exfor3/endf00.htm.

[17] Thermal neutron Capture Gammas by Target, NDS, IAEA, http://www-nds.iaea.org/oldwallet/tnc/ngtblcontentbyn.shtml.

[18] Hahn, O; Meitner, L. *Z. Phys.*, 1924, 29, 169.

[19] Pauli, HC; Alder, K; Steffen, RM. "The Theory of Internal Conversion," in *The Electromagnetic Interaction in Nuclear Spectroscopy*, edited by D. Hamilton, North-Holland, Amsterdam, 1975, Chap.10.

[20] Rosel, F; Fries, HM; Alder, K; Pauli, HC. "Internal Conversion Coefficients for all Atomic Shells." *Atomic Data and Nuclear Data Tables*, 1978, vol. 21, 91,

[21] Lee, MA; Nuclear data sheets for A=158, *Nuclear Data Sheets*, 1989, 56, 158.

[22] Bricc 2.0a. *Band-Raman International Conversion Coefficients*, BNL, http://www.nndc.bnl.gov/bricc/.

[23] Lederer, CM; Shirley, VS. *Table of Isotopes*, 7th Edition, Wiley, New York, 1978.

[24] Chen, MH; Crasemann, B; Mark, H. *Atomic Data and Nucl. Data Tables*, 1979, vol. 24, 13.

[25] Krause, MO. *J. Phys. Chem.*, Ref. Data, 1979, vol.8, 307.

[26] Dillman, LT. EDISTR - A Computer Program to Obtain a Nuclear Decay Data Base for Radiation Dosimetry, Oak Ridge National Lab. Report ORNL/TM-6689, 1980.
[27] Larkins, FB. *Atomic Data and Nucl.* Data Tables, 1977, vol. 20, 313.
[28] Bethe, HA; Ashkin, J. "Passage of radiations through matter." *Experimental Nuclear Physics*, Wiley, New York., 1953, vol.1, 166.
[29] Sternheimer, RM. "The density effect for the ionization loss in various materials." *Phys.Rev.*, 88, 851, 1952.
[30] Uehling, EA. "Penetration of heavy charged particles in matter." *Annual Rev. Nucl. Sci.*, 1954, No 4, 315.
[31] Conen, ER; Taylor, BN. "The 1973 least-squares adjustment of the fundamental constants." *J. Phys. Chem. Ref. Data*, 1973, No 2, 663.
[32] Moller, C. "Zur Theorie des Durchgangs schneller Elektronen durch Materiie." *Ann. Phys.*, 1932, vol. 14, 568.
[33] Rohrlich, F; Carlson, BC. "Positron-electron differences in energy loss and multiple scattering." *Phys. Rev.*, 1953, vol. 93, 38.
[34] Inokuti, M. Inelastic collisions of fast charged particles with atoms and molecules- the Bethe theory revisited, *Rev. Mod. Phys.*, 1971, vol.43, 297.
[35] Bichsel, H. *Passage of charged particles through matter*, in American Institute of Physics Handbook, 3rd Edition, D.E.Gray ed., New York, 1972, 8.
[36] Walske, MC. Stopping power of L-electrons. *Phys. Rev.*, 1956, vol. 101, 940.
[37] International Commision on Radiation Units and Measurements. *Stopping Powers for Electrons and Positrons*, 1984, ICRU Re 37.
[38] Bruckner, G; Czermak, A; Rauch, H; Welhammer, P. Position sensitive detection of thermal neutrons with solid state detectors (Gd Si planar detectors). *Nuclear Instruments and Methods in Physics Research*, 1999, A 424, 183.
[39] Murin, AN. *Physics Principal of Radiochemistry*, Moscow, Higher School, 1971.
[40] Abdushukurov, DA; Bondarenko, DV; Muminov, Kh.Kh; Chistyakov, DYu. *Calculations of the Efficiency of Registration of Thermal Neutrons by Complex Converters Constructed on the Basis of Gadolinium Foils.* Archives of Los Alamos National Laboratory USA, 2007, 1-22. http://xxx.lanl.gov/ftp/arxiv/papers/0711/0711.1282.pdf.
[41] X-RAY DATA BOOKLET, *Center for X-ray Optics and Advanced Light Source*. LBNL, USA, http://ie.lbl.gov/atomic/x2.pdf/.

[42] Cohen, ER; Taylor, BN. The 1986 *Adjustment of the Fundamental Physical Constants*. CODATA Bulletin 63 (values republished most recently in Physics Today, August 1997, BG7-BG11). 1986.
[43] X-Ray Mass Attenuation Coefficients http:// physics. nist. gov/ PhysRefData/ XrayMassCoef/tab3.html.
[44] Gebauer, B; Alimov, SS; Klimov, A.Yu. et al. Development of hybrid low-pressure MSGC neutron detectors. *Nuclear Instruments and Methods in Physics Research*, 2004, A 529, 358-364.
[45] Breskin, A; Charpak, G; Majewski, S. On the low pressure operation of multistep avalanche chambers. *Nuclear Instruments and Methods in Physics Research*, 1984, 220, 349.
[46] Abdushukurov, DA; Zanevsky, Yu.V; Movchan, SA; et al. "Multiwire Low Pressure Chamber with a High Gas Amplification Coefficient." *Instruments and Experimental Techniques*, 1983, v. 26, 1287.
[47] Dangendorf, A; Demian, H; Friedrich, et al. Thermal neutron imaging detectors combining novel composite foil convertors and gaseous electron multipliers. *Nuclear Instruments and Methods in Physics Research*, 1994, A 350, 503-510.
[48] Abbrescia, M; Iaselli, G; Mongelli, T; et al. Resistive Plate Chambers with Gd-coated electrodes as thermal neutron detectors. *Nuclear Instruments and Methods in Physics Research*, 2004, A 533, 149-153.
[49] Kenji Takahashi, Seiji Tazaki, Junji Miyaharaa, et al. Imaging performance of imaging plate neutron detectors. *Nuclear Instruments and Methods in Physics Research*, 1996, A 377, 119-122.
[50] Gunji, S; Yamashita, Y; Sakurai, H; et al. Development of Neutron Imaging Detector Using Capillary Phenomena and Liquid Scintillator, 2005. *IEEE Nuclear Science Symposium Conference Record*, 2899-2902.
[51] Gilvin, PJ; Matheison, E; Smith, GC. Position resolution in Multiwire Chamber with Graded-Density Cathodes. *IEEE Trans., Nucl. Sci.*, 1981, v. NS-28, No 1, 835.
[52] Abdushukurov, DA; Abduvokhidov, MA; Bondarenko, DV. "Spatial Resolution of the Multistep Avalanche Chambers." *Instruments and Experimental Techniques*, 2007, Vol. 50, No. 3, 333-335.
[53] Abdushukurov, DA; Zanevskii, Yu.V; Peshekhonov, VD. "Effect of gas-mixture composition on characteristics of multistage cascade chambers." *Instruments and experimental techniques*, 1989,Vol. 32, No 1, pt 1, 78-81.

[54] Abdushukurov, DA; Abduvaliev, AA; Abduvokhidov, MA; Muminov, Kh.Kh. "Structural Features of the Multistep Avalanche Chambers." *Instruments and Experimental Techniques*, 2007, Vol. 50, No. 1, 37-40.

[55] Breskin, A; Charpak, G; Sauli, F. *A Multistep Parallax Free Imaging Counter*, CERN-EP/81/106, CERN, 1981.

INDEX

A

absorption, vii, 22, 24, 26, 27, 40, 42, 44, 46, 47, 49, 52, 68, 73, 77, 104, 108, 111, 112
acetone, 76, 78, 79, 84, 85, 86, 87, 88
additives, 75, 87, 88, 114
adjustment, 22, 117
aluminum, 94
amplitude, 72, 74, 81, 82, 83, 84, 114
applications, 4, 10, 72, 101, 102
argon, 76, 78, 79, 81, 84, 85, 87, 88, 114
atmospheric pressure, 67, 71, 72
atomic nucleus, 10, 17
atoms, 17, 18, 19, 75, 108, 111, 117
authors, 2, 34, 94, 96, 105, 110
Avogadro number, 18

B

background, 1, 29, 30, 31, 88, 96, 97, 101, 105
background radiation, 96
backscattering, 95
beams, 93, 96
benzene, 106
beryllium, 90
binding, 110
binding energies, 110
breakdown, 89, 90

C

capillary, 4, 73, 105, 106, 113
CERN, 119
channels, 83, 108
collisions, 4, 16, 17, 18, 19, 20, 111, 117
components, 16, 18, 21, 24, 76, 93, 111
composition, 5, 23, 73, 101, 118
compounds, 43
concentration, 84, 85, 86, 87, 88, 104, 114
conductor, 47, 113
configuration, 4, 102, 104
construction, 10, 73, 107, 108, 109
contamination, 101
control, 83, 87, 94, 102
conversion, vii, 2, 3, 5, 9, 10, 11, 12, 13, 15, 24, 36, 40, 47, 48, 54, 67, 71, 72, 73, 75, 76, 79, 80, 81, 83, 88, 96, 102, 103, 104, 107, 108, 109, 111, 112, 114
copper, 99

D

database, 12, 24
decay, 10, 11, 29, 31, 109
density, 6, 17, 18, 21, 22, 24, 29, 42, 43, 71, 81, 94, 117
deposition, 97, 99
detection, 1, 4, 28, 71, 72, 75, 81, 82, 90, 94, 95, 96, 97, 101, 102, 105, 107, 110, 115, 117

diffusion, 95, 98, 99
diodes, 107
discontinuity, 90
displacement, 87
distribution, 75, 87, 90, 108, 110
drawing, 35, 48, 50, 76, 107

E

electric field, 17, 62, 67, 71, 75, 76, 79, 100, 111, 113
electrodes, 72, 73, 75, 76, 88, 101, 102, 118
electromagnetic, 10, 11, 110
electronic circuits, 108
emission, 5, 6, 10, 12, 15, 16, 17, 24, 27, 28, 36, 37, 38, 41, 52, 53, 54, 59, 60, 61, 62, 72, 81, 93, 94, 96, 101, 111
energy transfer, 18, 19
environment, 16, 21
estimating, 4, 54, 84
ethanol, 106
excitation, 10, 15, 17, 19, 20, 41, 103, 104, 111
exposure, 30, 83, 104
extraction, 97, 98, 99
extrapolation, 35

F

fabrication, 99, 108
feedback, 78, 99
films, 48, 50, 52, 53, 103, 113
fission, 1, 29, 101
fluctuations, 93
foils, viii, 2, 3, 23, 25, 31, 38, 52, 62, 67, 68, 70, 71, 73, 97, 99, 107, 112, 113, 116
Ford, 115
formula, 17, 18, 20, 81, 91
fragments, 1

G

gadolinium, vii, viii, 2, 3, 6, 7, 8, 12, 15, 22, 23, 24, 25, 26, 27, 28, 29, 30, 31, 32, 33, 34, 35, 36, 37, 38, 39, 44, 46, 47, 48, 49, 50, 51, 52, 53, 54, 62, 68, 69, 70, 72, 75, 76, 82, 88, 101, 111, 112, 113, 116

gamma radiation, 40
gamma rays, 1, 10, 106
gases, 17, 51, 75, 76
granules, 101
grids, 88, 89, 97, 98
groups, 24, 50
growth, 52, 53, 71, 72, 94

H

half-life, 7
height, 72, 83, 94, 95
hemisphere, 24, 27, 28, 31, 38, 52, 53, 54, 73
histogram, 9, 13
hybrid, 4, 72, 113, 118
hydrogen, 21

I

ideal, 31, 91, 92
image, 73, 84, 102, 103, 104
incidence, 3, 13, 18, 19, 20, 21, 31, 82, 83
interaction, vii, 5, 6, 7, 24, 36, 43, 71, 83, 85, 101
interactions, 43, 102
ionization, 17, 21, 22, 23, 71, 72, 75, 76, 77, 79, 80, 81, 88, 94, 97, 110, 111, 117
ionizing radiation, 102
ions, 18, 99
irradiation, 15, 29, 30, 31, 34, 106
isobutane, 94, 97
isotope, vii, 2, 3, 5, 7, 8, 9, 26, 28, 31, 34, 36, 38, 39, 49, 50, 52, 54, 84, 111, 112

J

Japan, 103

L

leakage, 108
line, 8, 15, 36, 48, 82, 83, 90, 95
localization, 71, 72, 93, 95, 98
luminescence, 102

Index

M

maintenance, 87
majority, 17, 27
manufacturing, 48
micrometer, 72, 73
miniaturization, 4
modeling, 2, 3, 4, 12, 32, 114
molecules, 85, 117
momentum, 11
Moscow, 116, 117
multiplication, 72, 95, 97, 98, 99

N

neon, 88, 114
neutrons, vii, viii, 1-9, 12, 13, 23, 24, 26, 27, 30-36, 38, 41, 42, 46, 48, 49, 50, 52, 53, 54, 55, 56, 62, 68, 71, 72, 73, 75, 95, 96, 97, 101, 103-108, 110-113, 115-117
noise, 31, 73, 95, 103, 104, 108
nuclear charge, 22
nuclei, 1, 6, 36, 49
nucleus, vii, 2, 5, 6, 10, 36, 43, 49, 111

O

oil, 72, 101
optimism, 32
order, 1, 2, 10, 17, 21, 24, 27, 67, 74, 87, 94, 97, 99, 101, 103
organic compounds, 88
oxide thickness, 102

P

parallel, 23, 51, 71, 72, 75, 81, 90, 98, 108
parallelism, 90, 91
parameter, 21, 22, 110
parameters, vii, 2, 5, 10, 18, 21, 34, 75, 111, 116
parity, 10, 111
particles, vii, 1, 2, 4, 5, 16, 17, 18, 19, 20, 21, 29, 71, 72, 73, 74, 77, 79, 82, 83, 84, 88, 93, 101, 103, 104, 106, 107, 114, 117
performance, 93, 94, 101, 118

PET, 103
photons, 42, 88, 97
physics, 107, 116, 118
pitch, 89, 90, 99
polarization, 17
positron, 10, 40, 43
positrons, 16, 18
power, 16, 17, 24, 99, 111, 117
pressure, 2, 4, 34, 51, 72, 75, 81, 88, 93, 95, 96, 97, 98, 99, 113, 118
probability, vii, 4, 5, 6, 10, 12, 16, 22, 24, 31, 36, 72, 89, 96, 102, 109, 112, 113
production, 10, 43
project, 51, 101
properties, 1, 17, 73, 101
protons, 1, 18, 20
prototype, 73, 99, 107
pulse, 72, 94, 95

Q

quanta, viii, 8, 9, 10, 28, 29, 40, 42, 44, 45, 46, 48, 77-79, 81, 103, 104, 112

R

radiation, i, iii, vii, 1, 2, 4, 10, 12, 15, 17, 29, 35, 36, 40, 41, 71, 77, 82, 85, 110, 111, 113, 114, 116, 117
radiography, 73, 93, 102, 103
radius, 19, 24, 81
range, 1, 6, 35, 45, 46, 51, 67, 71, 72, 81, 82, 84, 85, 88, 89, 102, 107, 108
reactions, 1, 103, 107
reception, 72, 113
recommendations, iv
region, 7, 95, 98, 108
resolution, vii, 2, 3, 4, 34, 51, 67, 71, 72, 73, 74, 75, 81-84, 87, 88, 90, 93, 96, 99, 105, 106, 107, 108, 110, 114, 118

S

saturation, 95
scattering, 43, 93, 98, 117
SCP, 98
selecting, 83

semiconductor, 2
sensitivity, 1, 34, 71, 72, 97
signals, 77, 79, 82, 83, 95, 102, 114
signal-to-noise ratio, 99
silicon, 4, 73, 107, 108, 113
solid state, 117
space, 34, 71, 77, 82
spatial frequency, 103
species, 29
spectrum, 10, 29, 83, 84, 95, 102
speed, 16, 19, 21
speed of light, 19
stability, 72, 87, 93, 95, 114
stabilization, 72, 87, 114
steel, 94, 99
strength, 76, 79, 97
substrates, 50
surface layer, 102
symbols, 11

T

Tajikistan, 74
temperature, 81, 87, 88, 99
thermal deformation, 99
thermal evaporation, 99
thermalization, 51
threshold, 17, 18, 97, 98
time resolution, 4, 75, 79, 86, 87, 88, 93, 96, 108, 110, 114
tracks, 16, 83, 90
trajectory, 77, 99
transition, 10, 11, 16, 36, 75, 88

transitions, 10, 85, 111
transmission, 94, 102
transmission electron microscopy, 102
transparency, 62, 69, 70, 76, 88, 113
tungsten, 90

U

uniform, 89, 90, 106
UV light, 95

V

vacancies, 4, 36, 40, 42
vacuum, 94
valence, 12
vapor, 88
variables, vii, 5
velocity, 75, 101

W

wavelengths, 6, 36, 38, 50, 52, 53, 73, 96, 110, 112
wires, 88, 89, 90, 95
WWW, 116

X

X-ray diffraction, 102

Y

Y-axis, 90